EIDELBERG SCIENCE LIBRARY | Volume 12

Hermann Dertinger · Horst Jung

Molecular
Radiation Biology

The Action of Ionizing Radiation
on Elementary Biological Objects

Translated by R. P. O. Hüber and P. A. Gresham

With a Preface by K. G. Zimmer

With 116 Figures

Springer-Verlag New York · Heidelberg · Berlin 1970

Dr. H. Dertinger, Priv.-Doz. Dr. H. Jung,
Prof. Dr. K. G. Zimmer

University of Heidelberg and Institute for Radiation Biology,
Nuclear Science Center, Karlsruhe/Germany

First published in 1969 | Heidelberger Taschenbücher
Band 57/58
„Molekulare Strahlenbiologie"
© by Springer-Verlag Berlin · Heidelberg 1969

ISBN-13: 978-0-387-90013-1   e-ISBN-13: 978-1-4684-6247-0
DOI: 10.1007/978-1-4684-6247-0

Preface

There can hardly be any doubt that radiation will continue to be an important factor in our lives. Present and future advances in atomic technology urgently require further work on research and development in the field of radiation biology if the maximum benefit is to be obtained at minimal risk from the various kinds of radiation that form a major by-product of nuclear processes. Consequently, it is also necessary to prepare students and younger scientists for doing such work.

The present book originates from teaching experience gained in lectures, seminars, and discussion groups started by the undersigned in 1957 and more recently held together with Drs. Dertinger and Jung. The friendly comments given to the German edition made us feel that it might be worthwhile to put the results of our efforts at the disposal of those to whom English is more familiar.

In agreement with the view, based on well-known facts, that most if not all of the more striking practical achievements have resulted from patient and careful investigations into some basic problem, the book aims at introducing the reader to the methods of thought and experiment used in molecular radiation biology as well as to the results obtained thereby. This approach to the general field appears to be indicated as at the present stage the major problems in radiation biology clearly require investigation at the molecular, i. e. physicochemical level. Whatever practical purpose one may have in mind (such as improvements in radiation therapy or reducing risks by accidental exposure): careful studies of the underlying basic mechanisms are the fastest and safest, in fact, the only way leading to useful results.

Karlsruhe, January 1970   K. G. Zimmer

Contents

Chapter 1. Introduction

Experience has shown that an introductory survey of the field is a good point of departure for a course of lectures. As radiation biology is an interdisciplinary subject, a careful consideration of its meaning and scope is, therefore, even more important than in other fields. Chemistry, for example, is clearly defined as the science of elements and their laws of combination and behaviour. A student of this subject acquires fairly clear ideas of the nature of chemistry, and these require few alterations later on. This is not the case with radiation biology, although at first sight it may seem clearly defined as *the science of the biological action of radiation*. The teaching of this subject is also different; like many other interdisciplinary subjects, it cannot be studied for a first degree. Students entering the field of radiation biology after the completion of their first degree find its complexity rather baffling at first; this applies even to pure science students. A physicist, for example, having been taught in a systematic manner, will now find that in radiation biology there are few standard textbooks to help him to become familiar with the subject.

An understanding of the problems posed by radiation biology is further complicated by the fact that experiments in this field depend on a wide variety of factors, even more so than in normal biological experiments. It is therefore possible, and even understandable, that interpretations and explanations can apparently be disproved, as the same experiment carried out under slightly different conditions can lead to an entirely different or even opposite set of results. This makes it difficult to recognize and choose those experiments which are important to radiation biology. In spite of these problems radiation biology is of special, one might almost say "existential", interest.

The rapid development of nuclear research and technology, as well as the wide range of applications of radiations in the treatment of disease, in widely differing industrial production processes, for the sterilization of drugs and medical appliances, for experiments in the food industry, and also the use of nuclear reactors and radioactive isotopes, make it essential to elucidate the mechanisms by which radiations act. This is the only way in which we are likely to learn "to live successfully with ionizing radiations" (Zimmer, 1968).

1

1.1. Historical Survey

The fact that there is no living organism, including man, that is unaffected by energetic nuclear radiations, means that any presentation of radiation biology must include large sections on biology and medicine. The inclusion of medicine adds considerably to the complexity of research in radiation biology. Furthermore, physics plays an important part in investigations into the nature of the basic interaction processes by which radiations cause damage in matter. To an outsider, radiation biology therefore gives the impression of a complex and almost impenetrable mixture of physics, chemistry, biology and medicine.

A rough division of the subject may be obtained by first of all separating medical radiation biology from the more basic research. The most important subdivisions of medical radiation biology are those of radiation patho-physiology and radiotherapeutic research. The development of drugs for the treatment of radioisotope poisoning, caused by the incorporation of radioactive materials, also belongs in the field of medical radiation biology (cf. Catsch, 1968).

Basic research in radiation biology includes the classical biological fields such as radiation cytology, radiation genetics and radioecology, as well as the modern subjects of radiation microbiology and molecular radiation biology.

Transitions between the different subjects are gradual and continuous, as there are many links between them; thus the divisions that have been made can only be considered as a first approximation. This is, however, quite adequate for our purposes, as we are mainly concerned with the definition of the position of molecular radiation biology within the whole field.

In this context, it is interesting to consider some of the historical aspects of the development of basic research in radiation biology, as this will also add considerably to our understanding of the problems. As with most new lines of research, its development was initiated by some quite accidental observations once the scene had been set by the discovery of ionizing radiations. A good example is that of Becquerel, who absent-mindedly carried a radium preparation around in his waistcoat pocket; this caused an inflammation of the skin, which he found healed only with difficulty. After the discovery of X-rays there were many scientists who, marvelling at their penetrating power, never tired of looking at images of the skeleton of their own hands. However, their enjoyment was soon dampened by the observation of peculiar changes in the exposed skin. Such phenomena stimulated interest in the action of ionizing radiations.

A characteristic of the period (that of *qualitative radiation biology*) which followed these initial observations was the emphasis on morphological investigations, which are popular even today when interest is apparently increasing once again. It was soon recognized that the reproductive and haemopoietic tissues are particularly sensitive to ionizing radiations.

It is, however, remarkable that while X-rays were used in the treatment of skin cancer as early as 1899, the discovery of the induction of skin cancer by X-rays was not made until 1902.

Progress in physics, chemistry and biology led to the second phase in radiation biology, which developed in the 1920s and is best described by the term *quantitative radiation biology*. This period is characterized by the application of mathematical and statistical methods to the interpretation of results (Blau and Altenburger, 1922; Dessauer, 1922). In a greatly simplified manner, this approach can be described as follows: the action of radiation is studied as a function of the absorbed radiation energy, i. e. as a function of dose. From a statistical analysis of the shape of the dose-response curves obtained (see Chapter 1.2), attempts are made to draw conclusions about the nature of the effective mechanisms. This approach led to the formulation of the "Hit Theory", which gradually, as the physical processes of radiation absorption have become better understood, has been greatly extended to allow for the complexity of experimental observations to what is now known as the "Target Theory". Even nowadays valuable information is gained from the analysis of dose-response curves. A peak of interest was reached around 1946—1947, when the books of Lea (1946) and Timoféeff-Ressovsky and Zimmer (1947) were published, finally establishing radiation biology as an independent branch of science. These two books had derived initially from common discussions; however, because of the events of World War II, their later developments had to be continued independently.

Although quantitative radiation biology has extended our knowledge of the effects of ionizing radiations considerably (for example, one could think of the marvellous successes of classical radiation genetics), it finally became apparent that this line of approach was not capable of leading to a complete understanding of the reaction steps intervening between the absorption of radiation energy and the biological endeffects. Nevertheless, the concepts of quantitative radiation biology have continued to be used, leading to investigations of the influence of various parameters which affect the extent and type of radiation damage. One could call this type of research the study of *modifying parameters in radiation biology*. This aspect of radiation biology began to receive a strong emphasis around 1945, although numerous investigations of this type had already been carried out. This type of experiment attempts, by modifying the external environment (by, for example, altering temperature and humidity, or adding substances that change the radiation sensitivity), as well as by using radiations of different qualities, to obtain results that allow a mosaic-like picture of the development of radiation damage to be composed.

The outlook, if one wishes to gain an understanding of the interactions of all of these factors, is not exactly promising. Efforts of this kind have led to a multitude of hypotheses, the number of which reaches the order of magnitude of the number of parameter combinations examined. This situation induced the well-known radiation biologist, Alexander Hollaender, to compare radiation biology with the history of a battlefield, on which all the battles were lost. This pessimism is probably caused not by the problems of radiation biology as such, but by its manner of evolution. Instead of studying individual reaction steps, all too often the endpoint of the development of radiation damage was examined, such as killing as a function of dose. To draw conclusions from this type of experiment as to the nature of the events leading up to the overall effect observed is a hopeless undertaking, even with the help of the most ambitious mathematical formalisms and the application of computers.

1.2. Dose-Response Curves and Special Aspects of Radiation Action

Even today, the dose-response curve is one of the most important diagrams in radiation biology. Depending on the type of experiment carried out greatly differing effects may be used as criteria of the actions of radiation: for example, the generation of free radicals, the inactivation of enzyme molecules, the loss of activity of deoxyribonucleic acid (DNA), the induction of a specific mutation, or the killing of a cell or an organism. In most cases the surviving fraction, e. g. the remaining relative enzyme activity etc., is plotted against the dose. When further reference is made to a dose-response curve or a dose-effect curve, it will mean this type of "survival curve". Other types of plot will be explained in each case.

It was soon recognized that these dose-response curves differed considerably from those obtained by the administration of various chemical reagents, e. g. poisons. This is clearly shown in Fig. 1. The existence of a threshold dose below which no adverse effects are observed is characteristic of the action of poisons. As the concentration is increased marginally above this threshold, the lethality increases rapidly. In contrast, the dose-effect curves of radiations increase only gradually with dose, and show some response even after very small doses. The gradient of the poison-response curve depends substantially on the variation of tolerance within the biological system studied, i. e. on the distribution of sensitivity to the poison. The smaller this variation, the steeper the gradient of the curve is expected to be. However, the dose-effect curve for radiation does not lend itself to interpretation in terms of "biological variability". The interpretation of survival curves will be dealt with in detail in Chapters 2 and 3. Although it can clearly be seen from Fig. 1 that the mechanisms of radiation damage and poisoning are different, one nevertheless has to make certain reservations; for example, if one plots the fraction surviving at a suitable concentration as a function of the exposure time, the action of certain chemical

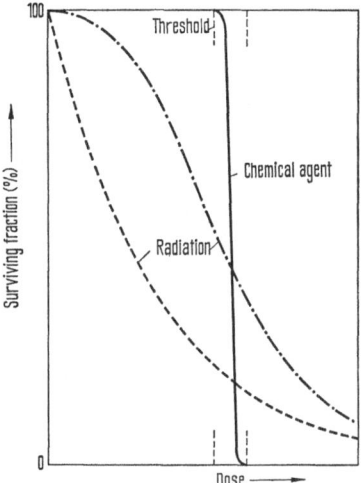

Fig. 1. Diagram illustrating dose-effect curves for the action of poisons and radiation. (Zimmer, 1961)

agents on DNA, as well as the influence of some antibiotics on bacteria, give dose-response curves similar to those produced by radiation.

It has already become apparent from the form of the dose-effect curve that there are distinctive features in the action of radiation, and this can be vividly demonstrated by what seems an almost trivial comparison. The small amount of energy consumed in a cup of hot tea is usually thought of as being both acceptable and beneficial. This does not apply, however, when the same amount of energy is taken up in the form of X-rays: in spite of the fact that this kind of energy transfer is not even noticed initially, it will lead after some hours or days to serious illness, or even to death. The problem demonstrated in this comparison represents the central aim of research in radiation biology, namely the elucidation of the mechanisms of the action of radiation: to elucidate the mechanism, in this context, means nothing more than to break down the complexities of radiation damage into terms of known physical and chemical processes.

1.3. The Temporal Stages of Radiation Action

When trying to follow the development of radiation damage in as much detail as possible, it is interesting and instructive to divide the complex chain of events that follow absorption of high-energy radiation in matter into three characteristic temporal stages (Platzman, 1958, 1962): during the first, or *physical stage* of radiation action, energy is transferred from the radiation to matter. This process leads mainly to molecular excitations

5

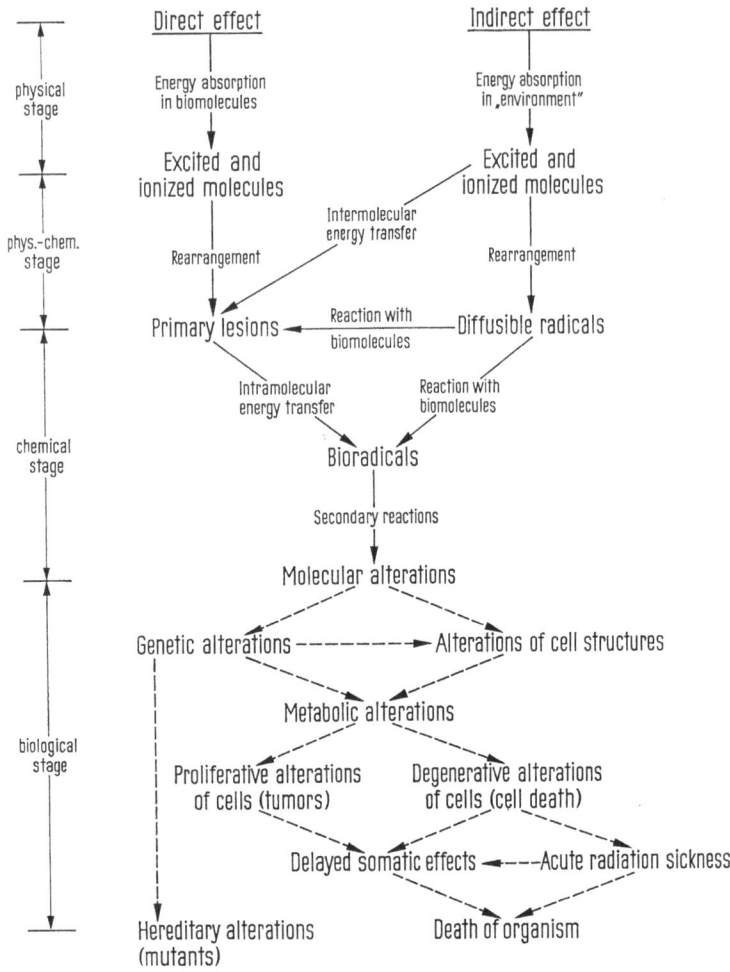

Fig. 2. The temporal stages of radiation action. (The reaction steps represented by broken lines are affected by metabolic processes.)

and ionizations in a drastically non-uniform spatial distribution. These primary species are usually extremely unstable and promptly undergo secondary reactions, either spontaneously or by collisions with molecules in their vicinity, to yield reactive secondary species. This second, or *physicochemical stage* may consist of a single reaction or a complex succession of reactions. Many of these interactions are not observed in other branches of physics and chemistry, although a few of them are known from photochemistry. The third, or *chemical stage* begins when the system finally reestablishes thermal equilibrium. In this phase the reactive species (usually free atoms or radicals) continue to react with each other and with their environment.

If the chain of events is initiated by the absorption of radiation energy in the system under investigation, for example in a DNA molecule or a particular biological structure, this is referred to as the direct action of radiation or the "direct effect" (Fig. 2). The primary processes of radiation absorption may, however, have occurred in the "environment" of a damaged biological molecule; this environment may include, for example, other biological molecules in the immediate vicinity. The energy absorbed by these molecules may be transferred to others by intermolecular energy transfer mechanisms, or alternatively by the liberation of diffusible radicals, such as hydrogen atoms, which may then react with undamaged biological molecules. Alternatively, if the biological molecules are in an aqueous environment, they may be attacked by diffusing reactive species (such as hydroxyl radicals, hydrogen atoms, or hydrated electrons) produced by the absorption of radiation by the water. The term "indirect effect" of radiation refers to both of these mechanisms (Fig. 2).

Regardless of their mode of formation, molecular changes occurring in a biological organism may cause alterations in the system which, in passing through the *biological stage,* finally lead to the development of the observed biological effect (amplification theory of organisms; cf. Jordan, 1948). During this stage of the development of radiation damage the metabolism of the affected organism is of particular importance (Fig. 2). The primary processes of radiation absorption merely cause small but significant injuries in the organism as a whole. However, the kind and the magnitude of the damage depend very much on whether the defect can be repaired, or whether the "machine", in attempting to operate under these perturbed conditions, tends to amplify the damage.

In spite of its limitations, this classification is of considerable assistance in any discussion of the complex succession of events following the absorption of energetic nuclear radiation. The orders of magnitude of the duration of the individual stages in an aqueous system have been estimated by Platzman (1962):

Physical stage:	10^{-13} sec
Physico-chemical stage:	10^{-10} sec
Chemical stage:	10^{-6} sec
Biological stage:	seconds to many years.

It is characteristic for these processes that each stage is short in comparison with the succeeding one; however the actual duration of a given stage depends very much on the system irradiated. In dry substances, for example, changes in the site of primary damage due to intramolecular energy transfer, and reactions of the biological radicals produced, may extend over minutes

7

or hours; this time may extend to days or even weeks if the irradiated system is kept at liquid nitrogen temperatures.

An ideal radiation biological investigation should lead to the elucidation of all the reaction steps in the various temporal stages shown in Fig. 2. It goes without comment that this aim has not yet been achieved. Nevertheless, it still remains the declared ultimate aim of research in radiation biology.

Fig. 2 gives some idea of the methods and techniques that can be used to follow the processes in the individual stages. Whilst the processes of the physical stage are virtually inaccessible to experimental investigation, and therefore have to be studied by means of physical measurements on model systems, there are methods available that can be used to obtain both qualitative and quantitative results in the physico-chemical stage. In this context, electron spin resonance (ESR) technique deserves a special mention. With the aid of this method, the production of primary radicals can be demonstrated; however, in order to avoid the occurrence of energy exchange, it is generally necessary to work at low temperatures. The existence of diffusible radicals, such as the hydrogen atoms, can also be demonstrated. In aqueous media, the rapid reactions between biological molecules and highly reactive radiolytic species are best investigated using pulse radiolysis techniques (cf. Ebert *et al.*, 1965). The secondary reactions of biological radicals in aqueous systems, finally, can be studied by the conventional techniques of analytical chemistry. For the investigation of the damaged molecule that is the endproduct of the chain of events, a large number of analytical procedures are available; these are based on differences in the physical and chemical properties between the damaged and the undamaged molecules, e. g. with respect to viscosity, sedimentation rate, optical absorption, solubility in acids etc. The individual procedures will be discussed later on.

1.4. The Significance of Molecular Radiation Biology

It can be seen from Fig. 2 that all reaction steps occurring later than the molecular changes are affected by the metabolism of the irradiated organism, so that the observed radiation effects depend on a multitude of complex biochemical reactions. Consequently, it hardly seems necessary to emphasize that the irradiation of an animal, such as a mouse, can never give a clear picture of the physical, chemical and biological processes which ultimately lead to the fatal break-down of the whole organism, as the test object "mouse" contains too many factors that interact in an unknown manner and are not accessible to measurement. In contrast, it might be expected that the analysis of reactions occurring at the molecular level is more likely to lead to the discovery of generally applicable principles. This view is supported by the results of modern biological research, which have shown that the differences between living organisms are less pronounced at the molecular than at the macroscopic (i. e. cellular and anatomical) level. This in turn is demonstrated by the existence of a universal genetic code and by the similarities between the basic components of all types of cells. It should therefore be possible, with a knowledge of radiation-induced molecular changes, to create a basis from which an understanding of the specific reactions of individual systems can be more readily obtained. Furthermore, it may be expected that a well founded knowledge of the molecular basis would render many experiments superfluous, as in some cases the results could be predicted; and this very ability to inductively forecast results represents an important step forward in the development of a scientific discipline. Accordingly, molecular radiation biology must be considered as a basic research discipline, which in itself may be taken as a justification of its existence. In addition, this line of research leads to a number of findings that are of practical importance. This is best illustrated by two examples: the sterilization of food and certain other items by irradiation prevents the multiplication of unwanted organisms, i. e. by the intentional destruction of their reproductive capacity. The basis for these procedures is an investigation of the radiation sensitivity of bacteria and a study of their ability to form radiation resistant mutants. The other example is taken from the field of radiotherapy. Considerable progress has been made, particularly in recent years, in the study and production of radiation sensitizing drugs, with the aid of which anoxic tumours can be made radiation sensitive. This allows the radiation dose to be reduced, which in turn leads to a reduction in the damage to healthy tissue in the vicinity of the tumour. The action of most of these drugs, which have been extensively studied using bacteria and bacteriophage, occurs at the genetic level.

Molecular radiation biology leads to directly applicable results whenever radiation is used to produce desired changes at the molecular and genetic level, for example, in the production of special mutants in seeds. At this time, as the complexities of biological processes are explained on the basis of biochemical reaction systems, it is probably reasonable to say that molecular radiation biology within the field of radiation biology as a whole, occupies a position similar to that of molecular biology in the field of biology.

1.5. An Introduction to Molecular Radiation Biology

From what has been said, it can already be seen that the main aims of molecular radiation biology are to examine the physical and chemical processes which lead to the damage in biological molecules, and to explain the failure of vital processes in terms of molecular changes. In a systematic treatment of the array of problems, a step-by-step discussion based on Fig. 2, in which each step would have to be explained experimentally and theoretically, would be particularly instructive. However, as has already been shown, radiation biology has not developed in this systematic manner. As a consequence, there is generally a lack of experiments with clearly defined aims, although there is usually no lack of results. A compromise is thus necessary between what is emphasized by the historical development of research in radiation biology on the one hand, and the type of approach indicated by our discussion of Fig. 2 on the other hand. Accordingly, the interpretation of dose-response curves will be dealt with first. In the following it will be necessary to consider the processes of energy transfer and energy absorption, which will at the same time be a more detailed description of the physical stage of radiation action. This will be followed by a discussion of the modern target theory, leading up to a discussion of the indirect action of radiation and an examination of the influences of various modifying parameters, such as oxygen concentration and temperature. Up to that point, the main concern will be the development of the mathematical, physical and chemical foundations of radiation effects; this covers the processes leading up to the biological stage and will be followed by a discussion of radiation effects in several particularly interesting and important test systems. It seems logical to begin with molecular systems, limiting the discussion to a consideration of two of the most important types of molecules that are essential for the continuation of metabolic processes: namely nucleic acids and enzymes. An attempt is made to construct and discuss specific models of enzyme inactivation, but in the case of nucleic acids there is not merely the question of the nature of the physico-chemical changes induced, but also that of correlating these, as far as possible, with the inactivation of the various nucleic acid functions. Viruses, occupying an important intermediate position between biological molecules and autonomous unicellular organisms, are extremely useful objects from the experimental point of view, as their genetic material consists of one single macromolecule, thus allowing rather small changes to be detected. The chapter about viruses will be followed by a discussion of bacteria, as representative of the simplest autonomous organisms that satisfy all the criteria of life, from reproduction to differentiated synthetic abilities. The last part of the book, as a projection to higher forms of life, is devoted to a discussion of the relationships between radiation sensitivity and biological complexity. Although mainly concerned with the action of ionizing radiations, occasional references will be made to experiments using ultraviolet light, when these help to bring about a better understanding of the processes of ionizing radiation. The investigation of the action of ultraviolet radiation does, however, belong primarily to the field of *Photobiology* (Smith and Hanawalt, 1969).

It will be obvious from this short outline of the scope of this treatise, that this is not an attempt to write a textbook of radiation biology. This would not, considering the present state of development of this field, be possible; on the other hand it is not necessary either, as no-one would seriously attempt to "learn" radiation biology as such. Instead some of the problems of molecular radiation biology, and the most important experimental and theoretical pathways which may lead to their solution, are presented. We are of the opinion that a fundamental and systematic presentation of a limited number of *typical* experiments serves this purpose far better than a treatise which attempts to be comprehensive.

Although no profound knowledge of the pure sciences is required in order to be able to follow the arguments, it is not possible to dispense altogether with mathematical, physical and chemical formalisms in the discussion of individual problems. This book is therefore aimed primarily at the science student, in the hope that it may, at some time, attract him towards this field, and also ease his introduction to the subject. The extensive listing of references will also enable the experienced worker in this field to extend his reading beyond the scope of the individual chapters. In addition, scientists in related fields should find it possible with the help of this book, to learn about the scope and problems of molecular radiation biology. The interested layman who has a basic knowledge of the pure sciences has also been borne in mind, and should not find it too difficult to follow this treatment of the subject.

References

Blau, M., Altenburger, K.: Z. Physik 12, 315 (1922).
Catsch, A.: Dekorporierung radioaktiver und stabiler Metallionen — Therapeutische Grundlagen. München: Thiemig 1968.
Dessauer, F.: Z. Physik 12, 38 (1922).
Ebert, M., Keene, I. P., Swallow, A. J., Baxendale, J. H. (eds.): Pulse radiolysis. New York-London: Academic Press 1965.
Jordan, P.: Das Bild der modernen Physik. Hamburg: Stromverlag 1948.
Lea, D. E.: Actions of radiations on living cells. Cambridge: University Press 1946.
Platzman, R. L.: In: Radiation biology and medicine. Ed.: W. D. Claus. Reading (Mass.): Addison-Wesley Press 1958, p. 15.
— Vortex 23, 372 (1962).
Smith, K. C., Hanawalt, P. C.: Molecular Photobiology. New York-London: Academic Press 1969.
Timoféeff-Ressovsky, N. W., Zimmer, K. G.: Biophysik I: Das Trefferprinzip in der Biologie. Leipzig: Hirzel 1947.
Zimmer, K. G.: Studies on quantitative radiation biology. Edinburgh-London: Oliver & Boyd 1961.
— In: Forschungspolitik, Heft 4. Ed.: Bundesminister für wissenschaftliche Forschung. München: Gersbach & Sohn 1968, p. 12.

Chapter 2. The Hit Theory

2.1. Basic Concepts

The hit theory is the oldest and at the same time the most illustrative of the theories that have been developed to interpret radiation dose-response curves. From the comparison of the effects of poison and radiation carried out in the preceding chapter it is evident that the form of radiation dose-response curves cannot be explained in terms of biological variability alone. The consideration of this initially unexplainable fact led to an entirely new approach: the application of quantum-physical ideas to biological problems. This laid the foundation for an interpretation of dose-response curves in terms of the hit theory, based on two physical observations and one postulate:

1. Ionizing radiation transfers its energy in discrete packets.

2. The interactions (hits) are independent of each other and follow a Poisson distribution.

3. The response under investigation occurs if a specified target has received a defined number (n) of hits.

This target, having a volume v (in units of cm^3), may represent the size of a sensitive structural element of the irradiated object. However, more detailed information about the nature of v can only be obtained after the discussion of the processes by which energy is transferred from radiation to matter, i. e. after a precise definition of a hit (Chapter 5). In a formal mathematical treatment, v is just a parameter for the radiation sensitivity. For present purposes the dose will be expressed in terms of "hits per cm^3".

The aims of the hit theory are twofold: to provide a mathematical description of dose-response curves and conversely to allow the characteristic parameters, such as hit and target numbers, to be determined from a given curve.

2.2. Single and Multiple Hit Phenomena

Dose-response curves are obtained by plotting, against radiation dose, the number of those units remaining after irradiation that are still capable of performing the same function as unirradiated units. This diagram is customarily referred to as a survival curve. The construction of such survival curves using the assumptions given in points 1 to 3 presents little difficulty. As the product $v \cdot D$ represents the mean number of hits within the volume v after the dose D, the probability of the occurrence of exactly n hits is, therefore, given by the Poisson distribution as:

$$P(n) = \frac{(v D)^n \, e^{-vD}}{n!}. \tag{2.1}$$

If n hits are required for the inactivation of an individual object, then any object receiving $n-1$ or fewer hits will survive. Accordingly, the survival curve is obtained by taking the sum of all units that have received $0, 1, 2, \ldots, n-1$ hits:

$$N/N_0 = e^{-vD} \sum_{k=0}^{n-1} \frac{(v D)^k}{k!} \tag{2.2}$$

where N stands for the number of survivors, and N_0 for the total population prior to irradiation.

For the special case where the response is produced by a single hit, equation (2.2) becomes:

$$N/N_0 = e^{-vD}. \tag{2.3}$$

This exponential "single-hit curve" will repeatedly be encountered in later chapters. With a hit number n greater than one, the curves described by equation (2.2) become sigmoid. Due to the exponential factor in the equation (2.2), these multi-hit curves are commonly plotted on a semilogarithmic scale. Fig. 3 shows a series of survival curves obtained from equation (2.2) and plotted in this manner. The exponential single-hit curve

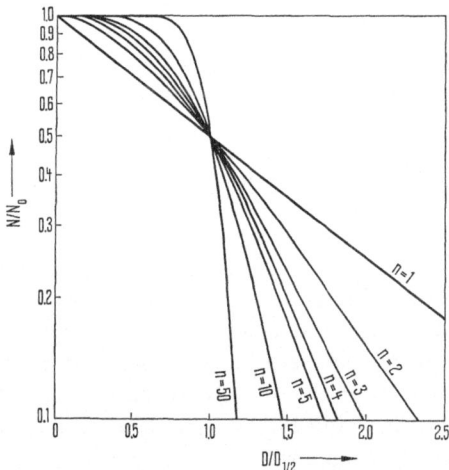

Fig. 3. Dose-effect curves as given by equation (2.2) for different hit numbers n in a semi-logarithmic plot. The curves are normalized to the "half-value-dose" $D_{1/2}$, i. e. to a surviving fraction of 0.5. (Zimmer, 1961)

13

$(n=1)$ is a straight line, while for higher hit numbers $(n=2, 3, \ldots)$ curves with increasingly pronounced shoulders are obtained.

How then, is it possible to solve the inverse problem: that of determining the values v and n from a given dose-response curve?

For a single-hit curve plotted on a semi-logarithmic scale ($\ln N/N_0 = -v D$), the magnitude of the target v is equal to the slope of the straight line. Thus, by specifying the dose, and measuring the corresponding survival rate, v is unambiguously determined. In practice, the dose at which 37% survival occurs (D_{37}), is found to be very useful, as at this point $N/N_0 = e^{-1}$, so that $vD = 1$, i. e. the target v is given directly by $v = 1/D_{37}$. The mean hit number $v \cdot D$ at the D_{37} is 1, which means that the total number of hits coincides with the number of objects irradiated. Consequently, the knowledge of the D_{37} (for example, in ergs/gram) can be used to obtain the dose in hits/gram, by dividing by the mean energy per hit. The fact that at the D_{37} the number of hits/gram is equal to the number of objects/gram will be used in equations (5.1) to (5.5).

A number of numerical and graphical procedures are available for the analysis of multi-hit curves; some of these are discussed in detail by Zimmer (1961). These procedures are, however, of theoretical rather than practical interest. In practice, n is best determined graphically, by plotting the experimental points together with a family of theoretical curves for different values of n (see Fig. 3), and looking for the curve that fits the experimental results most closely. It is necessary at this point to emphasize that even the best fit between experimental data and a theoretical curve is no proof that the number n obtained really represents the number of hits required to produce a response. The influence of many parameters must be investigated before such a number is unambiguously determined. It appears that in most cases so many factors may affect the determination of the hit-number, that it seems a fruitless task to carry it out.

2.3. Dose-Response Curves of Multiple Target Systems

The considerations will now be extended to objects having more than one sensitive structure. This corresponds to the assumptions that an object may have a number of formal targets, and that a response only occurs when each target has been hit a specified number of times. One example would be the killing of yeast colonies, each colony consisting of m cells, and only being killed when each of the m cells has received n hits. According to probability theory, the "killing curve" resulting from equation (2.2) must be taken to the mth power, giving:

$$N^+/N_0 = \left(1 - e^{-vD} \sum_{k=0}^{n-1} \frac{(v D)^k}{k!}\right)^m \tag{2.4}$$

14

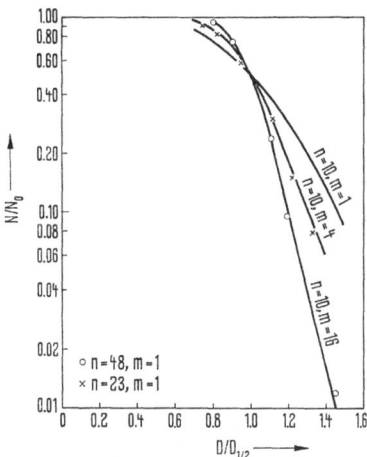

Fig. 4. Approximation of multi-hit curves with several targets by curves with only 1 target. (Glocker and Reuss, 1933)

where N^+ refers to the number of non-survivors. The validity of this expression can be extended to cover targets of varying magnitude v_i and different hit-numbers n_i:

where

$$N^+/N_0 = \prod_{i=1}^{m} (1 - B_i),$$

$$B_i = e^{-v_i D} \sum_{k=0}^{n_i-1} \frac{(v_i D)^k}{k!}.$$

(2.5)

Such complicated expressions, however, lose their practical significance because of the inherent inaccuracy of dose-response curves, preventing an unambiguous determination of the parameters v, n and m. An example will show clearly how difficult it is, in practice, to distinguish even between multi-hit curves for single and divided target regions. Fig. 4 shows three ten-hit curves, corresponding to 1, 4, or 16 targets in order of increasing slope. The experimental points plotted on this graph practically coincide with these curves over almost two decades, yet they were derived from equation (2.2) for simple multi-hit curves with $n = 23$ and 48. In the face of such difficulties, there seems to be no point in discussing theories that are even more complex. Readers interested in a more comprehensive treatment can find it in the work of Zimmer (1961).

Dose-response curves can be presented in a greatly simplified manner if only one hit is required for each target. The corresponding form is derived by a straight forward modification of (2.4):

$$N/N_0 = 1 - (1 - e^{-vD})^m = 1 - (1 - me^{-vD} + \ldots \pm e^{-mvD}).$$

(2.6)

15

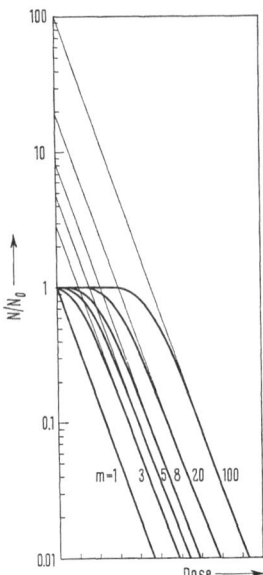

Fig. 5. Determination of the number of targets ("extrapolation number") for single-hit curves with m targets, according to equation (2.7). (Atwood and Norman, 1949)

At high doses, those terms following me^{-vd} can be neglected, leading to:

$$\ln N/N_0 = -v\,D + \ln m\,. \tag{2.7}$$

This means that the shoulder curve (2.6) plotted on a semi-logarithmic scale approaches an exponential dependence asymptotically, i.e. the linear part of equation (2.7), the slope of which equals v (Fig. 5). The point of interception obtained by extrapolation of equation (2.7) to $D=0$ yields the number of targets m: for this reason the parameter m is known as the extrapolation number.

All too easily, the hit-theory as developed up to this point may lead to far too rigid a view of the action of radiation. This is only partly a result of the use of terms such as "hit" for the critical absorption event which invite a comparison with "target-practice" on a rifle-range. The extensive oversimplification of the complex reactions undergone by an irradiated substance is a more serious consequence of the use of the hit theory. It is difficult to imagine that the $n-1$ hits occuring before the final hit, that leads to the observed response, will have absolutely no effect. In other words, sublethal damage would be expected to lead to sensitization, and therefore to uncertainty about the actual point of response. The rigorous treatment of such effects has to be reserved for the stochastic considerations of the next chapter. However, retrospective allowance can be made in the hit theory for this biological uncertainty.

2.4. The Influence of Biological Variability on the Form of Dose-Response Curves

The only possibility of adapting the hit theory to biological realities lies in a variation of the parameters v, n and m. The variation of v corresponds to a distribution of sensitivities within the irradiated population; a variation of n corresponds to the above-mentioned uncertainty in biological response due to the possible influence of sublethal hits, and a variation of the target number m could, for example, reflect different numbers of chromosomes. This allowance, however imperfect, made for the so-called biological variability by varying the parameters v, n, and m, will now be described and its consequences examined.

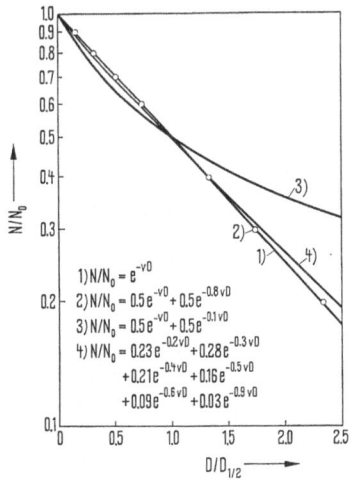

Fig. 6. Influence of variation of v on the form of single-hit curves. (Zimmer, 1941)

a) *Variation of* v. A variation in the size of the target has relatively little effect on the single-hit process. This is convincingly demonstrated in Fig. 6, which shows four single-hit curves for different values of v. Curve 1 is a pure single-hit curve, and the points lying on it (curve 2) refer to the situation in which there are two targets, one being 20% smaller than the other; the single-hit curve 4 has been calculated assuming a number of different values of v. However, a significant change is not obtained unless drastic differences are postulated, e. g. that half of the population have a target 1/10 of the size of the other half (curve 3). With variable targets, the slope of single-hit curves tends to decrease at high doses, as shown in Fig. 6. A simple interpretation of this is that at high doses the relatively insensitive units, i. e. those with a small target, determine the slope of the curve, while at low doses it is the more radiation sensitive

17

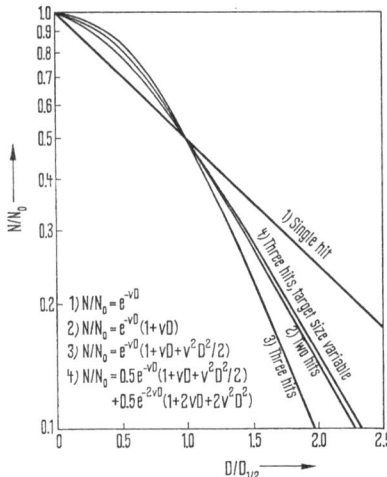

Fig. 7. Influence of variation of v on the form of multi-hit curves. (Zimmer, 1941)

units that are responsible for the steepness of the initial slope. However, as the dependence on v is not very strong, its variation will have little influence in practice, for biological variations between the systems under investigation tend to be small.

In contrast, the influence on a multi-hit curve of a variation in v is more pronounced, as shown in Fig. 7. In addition to normal single-, double- and treble-hit curves (curves 1, 2 and 3), a 3-hit curve is plotted (curve 4) reflecting the situation where one half of the objects have a target of v, and the other half a target of $2\,v$. Here too, as a result of biological variability, the slope of the curve decreases, thereby leading to estimates of hit number that are misleadingly low; this is shown by the fact that curve 4 practically coincides with a normal 2-hit curve.

b) *Variation of n and m.* These cases, which will not be dealt with in detail, also lead to a low estimate of hit number. When m is varied, it is often possible to describe the curves in simpler terms, i. e. by an expression with fewer targets and a corresponding change in the number of hits. Examples of this kind have been given by Zimmer (1961).

In investigations of the influence of biological variability, it is generally observed that the distortion of the curves increases with increasing hit number. This could be taken as meaning that dose-response curves for high hit-numbers reflect increasing biological variability rather than the Poisson statistics of the hits. This supposition turns out to be correct. To demonstrate this, it is assumed that biological variability follows a Gaussian distribution. If multi-hit processes are represented using a linear abscissa and a Gaussian ordinate (probability paper), then plots with upward curvature are obtained

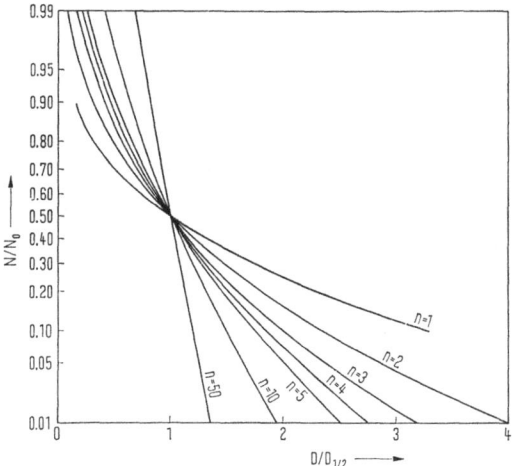

Fig. 8. Multi-hit curves on a probability plot: Abscissa on linear, ordinate on Gaussian scale. (Zimmer, 1961)

for low hit numbers, since the Poisson distribution does not yield a straight line on probability graphs. For the supposition to be correct, the curves should approach linearity for high hit-numbers. This is seen to be confirmed in Fig. 8.

Thus, a link with the poison-curve of Fig. 1 has been obtained. In the light of what has just been said it could be considered as a multi-hit curve, and the observation of a threshold could be explained by the pronounced shoulder of the semi-log curves having a very large hit number. However, the overall shape of the curve must be explained in the terms of biological variation, as was done in the previous discussion of Fig. 1.

2.5. The "Relative Steepness" of the Dose-Response Curve

In view of the fact that biological variability is unavoidable and that it influences the shape of dose-response curves, it could be asked if it would not be worthwhile to dispense with the Poisson distribution in the interpretations of dose-effect curves. Bearing this in mind, an attempt will now be made to generalize the hit theory, in order to obtain a better mathematical formulation of biological variation. To achieve this, the expression for the multi-hit curve is replaced by the generalized expression:

$$N/N_0 = 1 - W(D), \qquad (2.8)$$

where $W(D)$ represents, for the moment, the probability that after the delivery of a dose D to an individual object chosen at random from a biological population, the test effect has occurred. Because of the biological

19

variability, each individual of this population will have a different probability of showing the response.

$W(D)$ is therefore a distribution function of the "test dose" and increases continuously from 0 to 1 with increasing dose. $W(0) = 0$, i. e. at the start of irradiation, the probability of the test effect is zero; on the other hand $W(\infty) = 1$, which means that the survival rate is zero after an infinitely large dose. The hit events are implicitly contained in this expression, as each hit increases the probability $W(D)$. For the calculation of the relative steepness of the dose-response curve (2.8) the probability density w is used:

$$w(D) = \frac{dW(D)}{dD}. \tag{2.9}$$

From this expression, the first and second order moments are derived:

$$m_1 = \int_0^\infty D \, w(D) \, dD \quad \text{and} \quad m_2 = \int_0^\infty D^2 \, w(D) \, dD, \tag{2.10}$$

where m_1 and m_2 define the variance:

$$\sigma^2 = m_2 - m_1^2. \tag{2.11}$$

From the variance the relative steepness S is obtained:

$$S = \frac{m_1^2}{\sigma^2}. \tag{2.12}$$

S, by definition, is a positive quantity, and is 1 for exponential dose-response curves. For curves with higher hit numbers, S is larger than 1, and the following principle can be stated generally: if the dose-response curve has a relative steepness S, then $\bar{n}$, the mean number of hits necessary to produce the test effect, is greater than or equal to S, i. e.:

$$\bar{n} \geq S. \tag{2.13}$$

The difficult proof of this principle may be found in the book of Hug and Kellerer (1966) who applied it to the action of radiation. A link with conventional hit theory is readily obtained by introducing, in place of $W(D)$, the multi-hit expression:

$$W(D) = 1 - e^{-vD} \sum_{k=0}^{n-1} \frac{(vD)^k}{k!} = \int_0^\infty v \, e^{-v\delta} \frac{(v\delta)^{n-1}}{(n-1)!} \, d\delta. \tag{2.14}$$

As would be expected, the calculation of S gives in this case:

$$S = n. \tag{2.15}$$

The inequality (2.13) is the mathematical formulation of the fact already mentioned that biological variability leads to hit numbers which are too small. Since S can be determined, using equations (2.10), (2.11), and (2.12), by planimetry of the dose-response curve, it is possible to determine the least number of hits n necessary to produce a response.

2.6. Possibilities of Deception by Single-Hit Curves

From the modifications to dose-response curves, arising from biological variability, which have been considered up to now, it is clear that the practical application of the hit theory, and also of any equivalent theory, is seriously limited by such "interferences", as well as by the previously mentioned difficulties in distinguishing between multi-hit curves with single and multiple targets (see Fig. 7). Unfortunately biological variability is not even the sole source of interference. Further modifications are to be expected when time-dependent effects are included, i. e. the dependence of the test effect on the intensity of the radiation, and also when the ionization density is taken into consideration. The influence of these factors will be considered in the next chapter, together with a treatment of the kinetic model of radiation action which will allow a more readily understandable presentation of these infering parameters.

In view of this list of complexities, the single-hit curve must be considered as an honourable exception. For example, only drastic variations of v have any noticeable effect, while variations of n and m are not possible anyway. It is certain, therefore, that single-hit processes involving a single target will usually lead to an exponential dose-response curve. Is it, on the other hand, possible to conclude that an experimentally observed exponential curve is always caused by a single-hit process?

Basically this is possible, although in especially unfavourable circumstances a curve similar to that resulting from a single-hit process can be obtained by the superposition of multi-hit curves. Such circumstances could be encountered, for example, in a population containing different stages of development each having a different multi-hit response. This is the case in the induction of sex-linked lethal mutations in Drosophila melanogaster (Zimmer, 1943). This exponential dose-response curve which is often referred to as a classical and well established example of a single-hit curve, can be obtained even with radiations of different qualities; but it has proved on examination with modern breeding techniques to be the sum of several rather exotic dose-response curves (Traut, 1963; Zimmer, 1966). In Fig. 9, it is shown how well a single-hit curve can be simulated by the simultaneous occurrence of several double-hit curves; the heavily drawn line is the superposition of four double-hit curves:

$$N/N_0 = \frac{1}{4} \sum_{k=1}^{4} e^{-\frac{8}{2k-1} vD} \left(1 + \frac{8}{2k-1} v D \right). \tag{2.16}$$

The deviations from the single-hit curve are so slight that they cannot be determined experimentally (Dittrich, 1960).

It is clearly shown by these examples, together with the information given in Chapter 2.4, that a formal analysis of dose-response curves may lead to rather questionable results. To conclude on these grounds that the

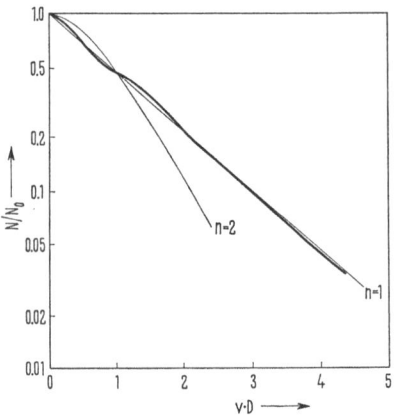

Fig. 9. Approximation of a single-hit curve by the superposition of 4 two-hit curves, following equation (2.16). (Dittrich, 1960)

hit theory is fundamentally "wrong", and should therefore be discarded, would be to ignore the real problem. The hit theory is no doubt basically a correct and consistent scheme. Although usually not applicable in its general form there remains the possibility that the analysis of dose-response curves may point to further useful experimental work. The heuristic aspect of the hit theory should not be underestimated. This is emphasized by the fact that its basic ideas concerning the existence of targets and hits, are undoubtedly correct, and this will be shown even more clearly in the treatment of the target theory given in Chapter 5. However — and this is one of the main reasons for its failure and, therefore, one of the main criticisms — most of the difficulties encountered in hit theory are caused, without doubt, by the rather "forced" postulate of a multi-hit action, whereby only the nth hit causes the test effect while all other hits have no effect. It is possible to make allowance for this fact in the kinetic model considered in the next chapter. However, one may well ask whether there actually is a multi-hit action as such? The results of the last few years suggest that in molecular radiation biology single-hit curves play a dominant role. This even extends, as will be shown in the next chapter, to occasional interpretations of curves having a shoulder purely in terms of "biological stochastics". In order to introduce this aspect into the classical hit theory, and furthermore to develop new ideas, the actions of radiation reflected by dose-response curves will in the following chapter be considered from a stochastical point of view.

References

Atwood, K. C., Norman, A.: Proc. nat. Acad. Sci. (Wash.) 35, 696 (1949).
Dittrich, W.: Z. Naturforschg. 15 b, 261 (1960).
Glocker, R., Reuss, A.: Strahlentherapie 46, 137 (1933).

Hug, O., Kellerer, A. M.: Stochastik der Strahlenwirkung. Berlin-Heidelberg-New York: Springer 1966.

Timoféeff-Ressovsky, N. W., Zimmer, K. G., Delbrück, M.: Nachr. Ges. Wiss. Göttingen VI, N. F., **1**, 189 (1935).

Traut, H.: In: Repair from genetic radiation damage. Ed.: F. H. Sobels. London: Pergamon Press 1963, p. 359.

Zimmer, K. G.: Biol. Zbl. **61**, 208 (1941).

— Physikal. Zschr. **44**, 233 (1943).

— Studies on quantitative radiation biology. Edinburgh-London: Oliver & Boyd 1961.

— In: Phage and the origins of molecular biology. Eds.: J. Cairns, G. S. Stent, and J. D. Watson. Cold Spring Harbor: Cold Spring Harbor Laboratory of Quantitative Biology 1966, p. 33.

Chapter 3. The Stochastics of the Action of Radiation

In this chapter, as already indicated by its title, an attempt is made to describe the action of radiation as a stochastic process, i. e. as a succession of random events. The process of refining the hit theory has led to the recognition of the fact that the initiation of a test effect is influenced not only by the hit number, but by a number of incidental events at the biological level.

The question of whether these processes caused by biological variation really are of a stochastical nature would, of course, require careful examination. However, stochastics of the vital processes give a model of the biological system showing a "dynamic instability", which is experimentally observable. The fact that synchronization of cell cultures can in general be maintained only for a few cycles, may be taken as a good example. This is due to initially insignificant variations in the biological parameters, which are then amplified by random changes in the physiological processes, and finally lead to a total loss of synchronization. The dynamic instability of a biological system is increased by the action of radiation; consequently a rigorous treatment of radiation action would have to consider a multiple stochastic model. Although the separate treatment of biological variability and radiation action limits a rigorous stochastical treatment, this simplification is necessary, because the mathematical formalisms would otherwise become unmanageable. This chapter essentially follows the treatment of the problem by Hug and Kellerer (1966); the interested reader who wishes to acquire a deeper understanding of the problems outlined here may refer to this book.

The stochastic considerations may be expressed by various mathematical formulations. Thus, the different sections of this chapter may give an impression of incoherence. In fact, an attempt is made to obtain a step-by-step transition (speaking in terms of hit theory) from multi-hit to single-hit processes: in the course of this presentation the stochastic idea will emerge in different mathematical formulations. Questions of dimensions will not be taken into consideration at this stage.

3.1. Kinetic Interpretation of the Dose-Response Curve

The kinetic formulation is particularly suitable for describing the biological stochastics of multi-hit processes. In contrast to the hit theory which is purely static in nature, the kinetic theory yields information about the velocity of change in an irradiated system. These changes occur in discrete steps; biological stochastics are taken into account by different probabilities of transfer from one level of damage to the next. This is obviously a logical way to allow for the fact, not expressed satisfactorily in the hit theory, that sublethal events influence the occurrence of the test effect. The kinetic description of a dose-response curve is based on the following assumptions: the system under consideration, e. g. a biological population, can be characterized by a series of states 0, 1, 2, ..., n (Fig. 10), the extent of

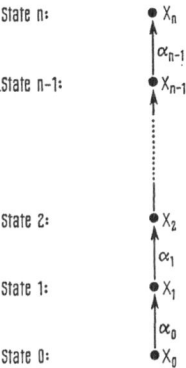

Fig. 10. Development of radiation damage from state 0 (beginning of irradiation) to state n (test-effect) with frequency numbers $x_0, ..., x_n$ and transition probabilities $\alpha_0, ..., \alpha_{n-1}$. (Hug and Kellerer, 1966)

damage increasing in this order and the test effect being triggered on reaching n. At any instant, the system is characterized by the "frequency numbers" of these states, i. e. by the relative number of units $x_0, x_1, ..., x_n$ in the states 0, 1, ..., n. The group of frequency numbers can be represented as a vector describing the state of the system:

$$\vec{x} = (x_0, x_1, ..., x_n) . \tag{3.1}$$

The extent of increase in radiation damage corresponds to transitions between states of damage, with probabilities $\alpha_0, \alpha_1, ..., \alpha_{n-1}$ (Fig. 10). Although nothing has been said about the significance of the individual states, a link with the hit theory can be made by identifying the components of the vector $\vec{x}$ with the relative number of objects; those without a hit are represented by x_0, those having one hit by x_1, etc. The changes in the system exposed to irradiation are taken as following a linear differential equation:

$$\frac{d\vec{x}}{dD} = A \vec{x} . \tag{3.2}$$

This expression implies that the velocity of change in an irradiated system is determined by the vector itself, and by a set of transition probabilities

α_0, α_1, ..., α_{n-1} that can be suitably represented by the matrix A, the so-called "transition matrix". A rigorous justification of equation (3.2) is derived from the theory of Markoff's chains (cf. Feller, 1957).

The system of linear differential equations (3.2) can be integrated in a formal way:

$$\vec{x} = e^{AD}\vec{x}_0 . \qquad (3.3)$$

where the exponential function is defined by its Taylor series:

$$e^{AD} = \sum_{k=0}^{\infty} \frac{A^k D^k}{k!} \qquad (3.4)$$

The initial state vector $\vec{x}_0$ characterizes the condition where all objects are in a state of "zero-hit", the frequency number of which is taken as 1, while all other states are not populated:

$$\vec{x}_0 = (1, 0, \ldots, 0) . \qquad (3.5)$$

At this point there is a possibility of taking into account the inhomogeneity of the biological material, by assuming that a fraction of the individuals are already in a higher state of damage prior to irradiation. This may be done by simply replacing the initial state vector (3.5) by some other suitable vector, which represents no problem when an analogue computer is used.

3.2. Multi-Hit Curves

How can dose-response curves be derived from the scheme developed in the previous section? To obtain the "relative number of survivors", all that is necessary is to sum the frequency numbers up to the state $n-1$, since the test effect occurs at state n:

$$N/N_0 = \sum_{k=0}^{n-1} x_k . \qquad (3.6)$$

Furthermore, if it is now assumed that the transition probabilities between the various states are identical ($\alpha_0 = \alpha_1 = \ldots = \alpha_{n-1} = \alpha$), then the matrix A, which corresponds to a multi-hit process, takes the following form due to the Poisson distribution:

$$A = \begin{vmatrix} -\alpha & 0 & 0 & \cdot & \cdot & 0 & 0 & 0 \\ \alpha & -\alpha & 0 & \cdot & \cdot & 0 & 0 & 0 \\ 0 & \alpha & -\alpha & \cdot & \cdot & 0 & 0 & 0 \\ \cdot & \cdot & \cdot & \cdot & \cdot & \cdot & \cdot & \cdot \\ \cdot & \cdot & \cdot & \cdot & \cdot & \cdot & \cdot & \cdot \\ 0 & 0 & 0 & \cdot & \cdot & \alpha & -\alpha & 0 \\ 0 & 0 & 0 & \cdot & \cdot & 0 & \alpha & -\alpha \end{vmatrix} \qquad (3.7)$$

Using this matrix, and applying the starting condition (3.5), the state vector is obtained:

$$\vec{x} = e^{-aD}\left(1, \alpha D, \frac{\alpha^2 D^2}{2!}, \ldots \frac{\alpha^n D^n}{n!}\right) . \qquad (3.8)$$

This gives the fraction of survivors, as defined by equation (3.6):

$$N/N_0 = e^{-aD} \sum_{k=0}^{n-1} \frac{(\alpha D)^k}{k!}, \qquad (3.9)$$

and this is the expression for the multi-hit curve already derived in the hit theory (equation 2.2), the only difference being that the transition probability has taken the place of volume v. However, this formal replacement applies only to this special case; in the more complex situations discussed later, the transition probability cannot be identified with the target volume.

The matrix (3.7) does not yet include the biological stochastics, which must now be introduced. All that has to be done is to consider the transition probabilities as being different, according to the conditions already laid down. The corresponding matrix, by an obvious generalization of (3.7), is:

$$A = \begin{vmatrix} -\alpha_0 & 0 & 0 & \cdot & \cdot & 0 & 0 & 0 \\ \alpha_0 & -\alpha_1 & 0 & \cdot & \cdot & 0 & 0 & 0 \\ 0 & \alpha_1 & -\alpha_2 & \cdot & \cdot & 0 & 0 & 0 \\ \cdot & \cdot & \cdot & \cdot & \cdot & \cdot & \cdot & \cdot \\ \cdot & \cdot & \cdot & \cdot & \cdot & \cdot & \cdot & \cdot \\ 0 & 0 & 0 & \cdot & \cdot & \alpha_{n-3} & -\alpha_{n-2} & 0 \\ 0 & 0 & 0 & \cdot & \cdot & 0 & \alpha_{n-2} & -\alpha_{n-1} \end{vmatrix} \qquad (3.10)$$

Some aspects of the dose-response curves derived from the matrix (3.10) are quite remarkable. It has been shown by Hug and Kellerer (1966) that they have a definite extrapolation number, on a semilogarithmic plot, i. e. that they approach exponential dependence asymptotically, even if only one transition probability differs from the rest. This is convincingly demonstrated in Fig. 11, where 5 "four-hit" curves, obtained using an analogue

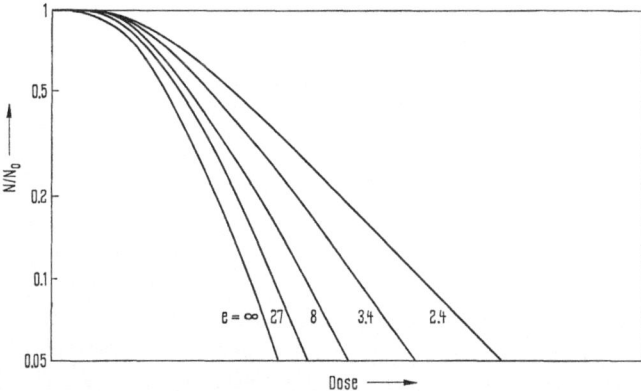

Fig. 11. Dose-effect curves in the kinetic model for $n = 4$. For equal transition probabilities α, the steepest curve obtained (a "4-hit" curve) has an infinite extrapolation number e. The other curves with $e = 27$, 8, 3.4, and 2.4 were calculated on the assumption that one of the probabilities α is smaller than the three others by the factors 0.66, 0.5, 0.33, and 0.25, respectively. (Hug and Kellerer, 1966)

computer, are shown together with the corresponding extrapolation numbers e. The slope of the curves decreases progressively as the difference between one transition probability and the other three increases; i. e. the extrapolation number becomes smaller. However, this extrapolation number by no means may be regarded as the number of targets. But this is only a slight disadvantage for practical purposes, as integer values are rarely obtained anyway. It is certainly misleading to refer to kinetic curves as "hit curves", as the term "hit" is not contained explicitly in the kinetic description. It may be better to refer to a "multi-step curve", particularly as this emphasizes that the kinetic curve generally differs from the hit curve.

3.3. Reverse Processes

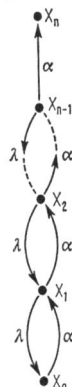

Biological systems often have the ability to eliminate radiation lesions. This process is described in the kinetic model by reverse transitions between various states of damage. The reaction scheme is shown in Fig. 12, where it has been assumed that the probability λ for the reverse process is the same for each state. The corresponding dose-effect curve is derived by splitting the transition matrix A into two parts, and entering these into equation (3.3):

$$A = A\,(\alpha_i) + A\,(\lambda), \qquad (3.11)$$

where $A\,(\alpha_i)$ is given by equation (3.10). The matrix $A\,(\lambda)$ contains the constant probability factor λ for the reverse transitions; the reverse trend is emphasized by using the form:

Fig. 12. Kinetic scheme of radiation action, taking reverse processes into consideration. (Hug and Kellerer, 1966)

$$A\,(\lambda) = \begin{vmatrix} 0 & \lambda & 0 & \cdot & \cdot & 0 & 0 & 0 \\ 0 & -\lambda & \lambda & \cdot & \cdot & 0 & 0 & 0 \\ 0 & 0 & -\lambda & \cdot & \cdot & 0 & 0 & 0 \\ \cdot & \cdot & \cdot & \cdot & \cdot & \cdot & \cdot & \cdot \\ \cdot & \cdot & \cdot & \cdot & \cdot & \cdot & \cdot & \cdot \\ 0 & 0 & 0 & \cdot & \cdot & -\lambda & \lambda & 0 \\ 0 & 0 & 0 & \cdot & \cdot & 0 & -\lambda & \lambda \\ 0 & 0 & 0 & \cdot & \cdot & 0 & 0 & -\lambda \end{vmatrix} \qquad (3.12)$$

The influence of reverse processes on a three-hit curve is shown in Fig. 13. The slope of the curve decreases as the relative importance of the reverse processes increases, and it gradually approaches an exponential form. The reverse processes often depend on the radiation intensity (or dose rate). Assuming, for example, the recovery parameter λ to be inversely proportional to the intensity, then an increase in intensity corresponds to a progression within the family of curves shown in Fig. 13, from the top down-

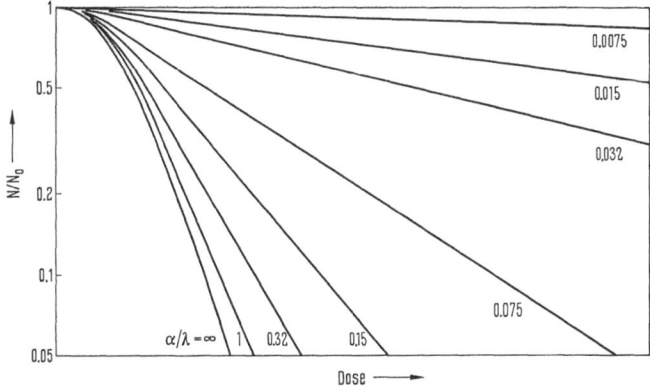

Fig. 13. Influence of reverse processes on "3-hit" curves. (Hug and Kellerer, 1966)

wards. This dependence on intensity is sometimes referred to as the "time-factor effect", which emphasizes that short-term irradiations at high intensity are often more effective than long-term irradiations at low intensity.

The kinetic model can also, in principle, be modified to allow for the influence of radiation quality on the form of dose-response curves. Higher ionization density is reflected in the action scheme of Fig. 10 by the possibility of transitions between non-adjacent stages. As a result, fewer elements of the transition matrix A are zero, and the dose-effect curves take forms corresponding to a lower-order kinetic, i. e. they tend to become purely exponential. A more detailed discussion is given by Hug and Kellerer (1966).

The kinetic model like the hit theory, has the disadvantage that various postulates about the nature of the mechanism of radiation action are inherent in the basic assumptions. The question of whether the shape of the dose-response curves really reflects a multi-hit process or is to be attributed to a high-order reaction kinetic still remains unanswered. These doubts seem to be particularly justified in the field of molecular radiation biology; consequently the possibility of obtaining a curved dose-response relationship without assuming a multi-hit process, has to be investigated.

3.4. A Formalistic Description of Dose-Response Curves

A formalistic mathematical description of curved dose-response relationships can be based on the fact that most of the experimentally observed curves approach exponential dependence at high doses. This may arise from biological stochastics alone, as was shown in the previous section. It, therefore, seems reasonable to commence with the expression:

$$\frac{d}{dD}\left(N/N_0\right) = -R\left(D\right) \cdot N/N_0 \,. \tag{3.13}$$

29

The quantity $R(D)$, which reflects the influence of biological stochastics, is referred to as "reactivity" by Hug and Kellerer (1966). As can be seen from equation (3.13), it is equal to the slope at the dose D of the dose-response curve plotted on a semi-logarithmic scale, and thus represents a "differential radiation sensitivity":

$$R(D) = -\frac{d}{dD}\ln(N/N_0).$$ (3.14)

If R is constant, this gives the exponential curve:

$$N/N_0 = e^{-RD}.$$ (3.15)

In the case of higher-order curves it is, in contrast, necessary to assume that reactivity increases with dose. The actual functional dependence has to be determined for each individual case. In principle, it is possible to describe a multi-hit process in this manner by suitable choice of the function $R(D)$; however, this leads to a complicated mathematical expression, which shows how artificial it is to assume a multi-hit process in this model. It seems more reasonable to describe shoulder curves by a function $R(D)$ that increases exponentially from a lower to a higher level with increasing dose:

$$R(D) = R_0 - R_1 e^{-\gamma D}.$$ (3.16)

The corresponding dose-response curves then take the form:

$$\ln(N/N_0) = -R_0 D + \frac{R_1}{\gamma}(1 - e^{-\gamma D}).$$ (3.17)

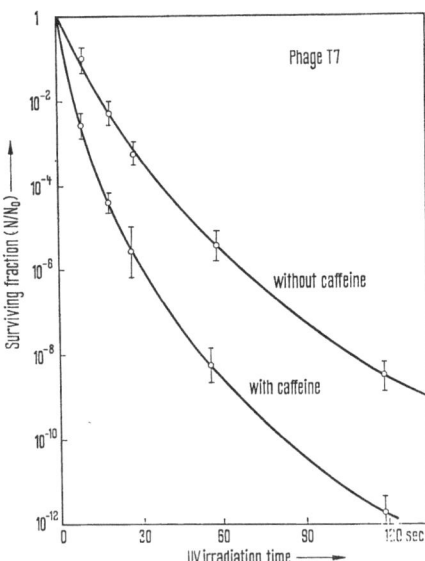

Fig. 14. UV inactivation of phage T7. The addition of caffeine prevents host cell reactivation. (Rontó *et al.*, 1967)

However, there are situations where the reactivity could possibly decrease exponentially with dose; this can be taken into account by reversing the sign of R_1 in equations (3.16) and (3.17), and it leads to curves having a smaller asymptotic slope. An experimental example of this is given in Fig. 14. The inactivation of phage T7 is plotted as a function of UV exposure time. The authors (Rontó et al., 1967) explain the decrease in sensitivity by a process of ultraviolet reactivation, i. e. absorption of a second UV-quantum resulting in partial elimination of radiation damage. Using probability theory, they arrived at an expression similar to equation (3.17) in which $R_0 = 0$, and R_1 is assumed to be negative, representing the probability that an undamaged phage will be injured by the absorption of a photon, and γ is defined as the reactivation probability per absorbed photon. The curve with the steeper slope (obtained in the presence of caffeine) has $R_1 = 2.5 \cdot 10^{-4}$ and $\gamma = 2 \cdot 10^{-5}$. This means that on average 1 in 4000 photons causes damage, while only 1 in 50,000 reactivates a damaged phage.

At this stage, it should be pointed out that the inactivation of the transforming ability and priming activity of DNA often yields dose-effect curves whose slopes decrease as the extent of inactivation increases. The question of whether these curves can be explained by a reduction in sensitivity with increasing dose, will not be dealt with until Chapter 11.

3.5. Dose-Response Curves of Colony Formation

The formal treatment of dose-response curves at least has the advantage that the parameters involved can be determined, as is clearly shown by Fig. 14. Nevertheless, it is not justified in all cases to explain shoulder curves on the assumption that reactivity increases with dose, or in other words, that resistance decreases. This final section will, therefore, concentrate on a point that has not been considered as yet: the test effect.

It is important, particularly when the ability to form colonies is tested, to enquire as to the real meaning of the corresponding dose-effect curves. The colony test is, of course, no direct measure of the radiation damage and, therefore, of the kinetics of radiation action, but reflects the consequences of a radiation lesion, i. e. the inability of cells to divide and to form a colony. Consequently, the corresponding dose-effect curves also depend on an important biological-stochastic factor: the biological "expectation value" of successful cell division after irradiation.

For the interpretation and mathematical description of colony curves, the probability that the irradiated cell will divide sufficiently often to produce a colony has to be calculated. The possibility of describing colony curves by an expression from the gambling theory ("gambler's ruin") has been discussed by Hug and Kellerer (1966). This leads to shoulder curves even though an exponential decrease with dose of the probability of division per cell has been assumed. It is certainly possible, using some other suitable statistical expression, to describe bacterial colony curves, which also usually have a shoulder. Under these circumstances the extrapolation number of these curves does not necessarily, as is often assumed, represent the number of targets in irradiated bacteria, since it could, for example, represent the critical number of daughter cells formed before failure to produce further cells (i. e. to form a colony), caused by sublethal damage, becomes apparent.

This concludes the discussion of the attempts to interpret dose-effect curves. It is felt that by describing them in detail, and as exactly as possible, various misconceptions and "new discoveries" might be avoided, and it is furthermore hoped that these considerations may have given some useful suggestions as to the manner in which the analysis of dose-response curves should be carried out in individual cases. Finally, it should be mentioned that there are special dose-effect curves for certain biological tests (such as the transforming ability and priming activity of DNA), the interpretation of which will be left to later chapters.

References

Feller, W.: An introduction to probability theory and its applications. Vol. I. New York: John Wiley & Sons 1957.

Hug, O., Kellerer, A. M.: Stochastik der Strahlenwirkung. Berlin-Heidelberg-New York: Springer 1966.

Rontó, G., Sarkadi, K., Tarjan, I.: Strahlentherapie **134**, 151 (1967).

Chapter 4. Primary Processes of Energy Absorption

Up to this point, the aim has been to describe the form of dose-response curves schematically, using the principles of hit theory, kinetic theory or general stochastics. A knowledge of the primary processes of absorption is necessary before this discussion can be extended. There are two ways to approach this problem: either by looking at the loss of radiation energy (the slowing down) of a charged particle, or alternatively by looking at the uptake of energy by molecules of the irradiated material. Both approaches will be used, commencing with a consideration of the processes of interaction of radiation with matter. This will include electromagnetic radiation, as well as charged and uncharged particles. The section is subdivided according to the different interaction processes.

4.1. X- and Gamma-Radiation

Whilst X-rays are produced by the slowing down of fast electrons in materials of high atomic number, γ-rays originate in the nucleus and often accompany the emission of α- and β-particles. However, they are both energetic electromagnetic radiations, differing only in their origin. An important property of these radiations is that their intensity decreases exponentially with depth of penetration x:

$$I(x) = I_0 \cdot e^{-\mu x}, \tag{4.1}$$

where I_0 is the intensity of the incident radiation and μ the attenuation coefficient of the irradiated material. The decrease in intensity can be caused by scattering of the quanta; in this process no energy is transferred, but the incident wave changes its direction. On the other hand it can be caused by a transfer of energy to the material. The attenuation coefficient μ consists therefore of a scattering coefficient σ and an absorption coefficient τ:

$$\mu = \sigma + \tau. \tag{4.2}$$

According to the Grothus-Draper principle only the absorbed radiation is important for the induction of a biological response, and not the scattered or transmitted radiation. Scattering of radiation is, therefore, excluded from this discussion, although it plays an important role in, for example, radiation dosimetry.

X- and γ-rays are absorbed by interactions with atomic electrons. There are three basic mechanisms:

a) Photoelectric Effect. In this process the incident quantum transfers all of its energy to an atomic electron. The kinetic energy of the electron, according to Einstein's equation, is equal to the quantum energy ($h\nu$) minus the binding energy A:

$$E_{kin} = h\nu - A. \tag{4.3}$$

The probability that this process will occur is highest when the energy of the quantum coincides with the binding energy of the electron emitted. This leads to sharp absorption edges such as the K-edge shown in Fig. 17. Irrespective of the electron shell concerned, a good approximation for the photo-absorption coefficient is:

$$\frac{\tau_{ph}}{\varrho} \sim \left(\frac{Z}{h\nu}\right)^3 (1 + 0.008 \cdot Z), \tag{4.4}$$

where Z is the atomic number of the irradiated element. The probability of the photoelectric effect increases rapidly with Z, and decreases with increasing quantum energy. In biological molecules, however, the atomic number has to be replaced by an "effective" value, the magnitude of which depends on the power to which Z is raised in the expression for the absorption coefficient. Table 1 gives the effective atomic numbers for the photoelectric effect in several biologically important materials. Other processes, such as pair production, have different values for Z_{eff}, as they have a different dependence on Z.

Table 1. *Effective atomic number for the photoelectric effect in some biologically important substances.* (Jaeger, 1959)

Material	Z_{eff}
Air	7.64
Water	7.42
Muscle	7.42
Bone	13.8
Fat	5.92

b) Compton Effect. In the Compton interaction, in contrast to the photoelectric effect, only a part of the energy of the incident quantum is transferred to the electron. This reduces the energy of the quantum, i. e. increases its wavelength. In addition the quantum changes its direction. The angle at which the electron is ejected depends on the amount of energy transferred. The distribution of energy between the quantum and the Compton electrons is shown in Fig. 15. Only 5% at the most of the energy of a 10 keV quantum can be transferred to an electron. Events in which

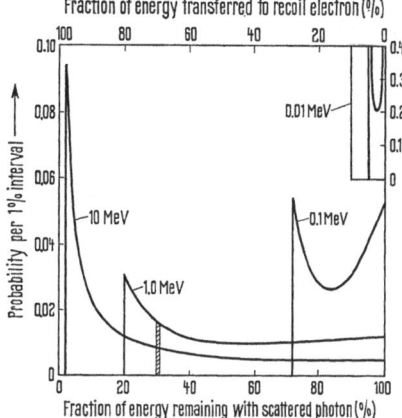

Fig. 15. Distribution of energy between scattered photon and emitted electron in a Compton interaction (White, 1951). Example: The probability that a 1 MeV photon retains 30 to 31 % of the total energy is 0.016 (shaded area)

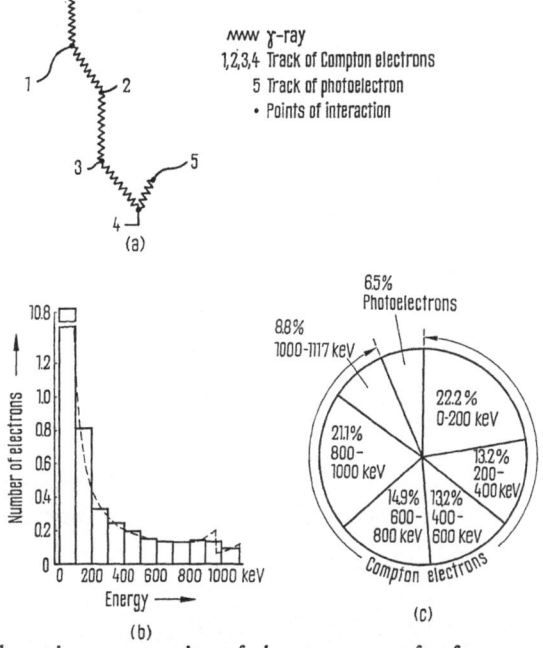

Fig. 16. a Schematic representation of the energy transfer from a γ-quantum by repeated Compton scattering followed by photoelectric effect. b Frequency distribution of the electrons released by ^{60}Co γ-radiation in water (average number per quantum). The broken line shows the spectrum of the Compton electrons. c Energy distribution of electrons released by ^{60}Co γ-radiation in water by Compton and photoelectric effect. (Spencer and Stinson, 1954)

either very little, or most of the energy is transferred to an electron have a high probability, so that the quantum is either scattered very little or by almost 180°. As the quantum energy increases, the probability of the transfer of large amounts of energy increases, giving a more and more pronounced "Compton edge". The probability for the Compton effect, which may be regarded as an inelastic scattering of photons, can be calculated using the Klein-Nishina formula, which is given in many textbooks on Atomic Physics.

Fig. 16 is a schematic representation of a ^{60}Co γ-quantum, which loses its energy step by step through Compton interactions, until finally all of the energy remaining is transferred to a photoelectron (a). It can be seen from the frequency distribution of Compton electrons shown in Fig. 16 (b), that most of the emitted electrons have energies between 0 and 100 keV, the frequency of more energetic electrons decreasing rapidly. The dotted line is the energy spectrum of the Compton electrons, with the Compton edge of the 1.17 MeV quanta of the ^{60}Co radiation at approximately 0.98 MeV and that of the 1.33 MeV quanta at 1.117 MeV. Fig. 16 (c) shows the proportion of the total energy absorbed that is transferred to Compton electrons within various energy intervals, and to photoelectrons. In this representation the Compton edges are reflected by the relative magnitude of the contributions in the intervals 800 to 1000 keV and 1000 to 1117 keV.

c) *Pair Production.* This interaction process becomes possible at quantum energies exceeding 1 MeV, and consists of the generation of an electron-positron pair, so to speak "from nothing", by an interaction with nuclear fields, or, using Dirac's formulation, from electron states of negative energy. The sum of the kinetic energies of the electron and the positron is equal to the energy of the quantum less twice the rest energy E_0 of an electron ($E_0 = m_0 c^2 = 0.51$ MeV):

$$E_{e^+} + E_{e^-} = h\nu - 1.02 \text{ MeV}. \tag{4.5}$$

The energy $h\nu - 2 m_0 c^2$ can be arbitrarily divided between the two particles; it does however depend on their angles of emission. Equation (4.5) shows that pair production can only occur at gamma-quantum energies greater than 1.02 MeV.

The relative frequency of these three processes depends on the nuclear charge of the irradiated material. The approximate relationships are:

$$
\begin{array}{ll}
\text{Photoelectric Effect} & \sim Z^4 \\
\text{Compton Effect} & \sim Z \\
\text{Pair Production} & \sim Z^2
\end{array}
$$

Furthermore, the frequency of the various interactions depends on the quantum energy. Although this will not be discussed in detail, it is interesting to compare the relative frequencies of photoelectric effect, Compton effect and pair production at different quantum energies. In substances of

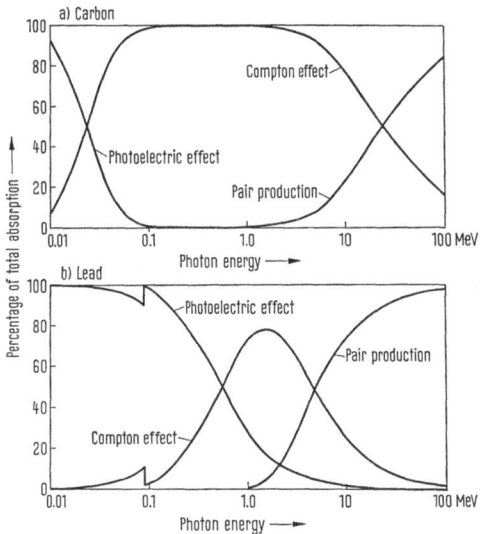

Fig. 17. Relative frequency of photoelectric effect, Compton effect, and pair production in carbon and lead. (White, 1951)

low atomic number, such as carbon and biological molecules in general, the dominant interaction process in the energy range 50 keV to 20 MeV is that of the Compton effect. Lead, having a high nuclear charge, gives rise to a much stronger photoelectric effect reflected by the appearance of the K-edge, and an increase in pair production (Fig. 17).

Half-value Layer. Electromagnetic radiation is attenuated exponentially with increasing depth of its penetration; a range can, therefore, not be defined. However, the "half-value layer" can be used as a measure of penetrating power. It is defined as that thickness of material which attenuates the intensity of the radiation by one-half. The half-value thickness for ^{60}Co γ-radiation is, for example, 11—12 cm in water or protein but only about 4 mm in lead.

4.2. Neutrons

High fluxes of neutrons can be obtained from nuclear reactors, and from a large number of nuclear reactions initiated by the irradiation of light nuclei with protons, deuterons or alpha particles. Even gamma-rays of sufficiently high energy can be used to generate neutrons (the photonuclear effect), although this method has no biological significance. It is possible nowadays to produce neutrons with energies in the range 10^{-2} to 10^8 eV. Only the biologically important interactions of neutrons with matter are of interest in the context of this book. The fact that the primary inter-

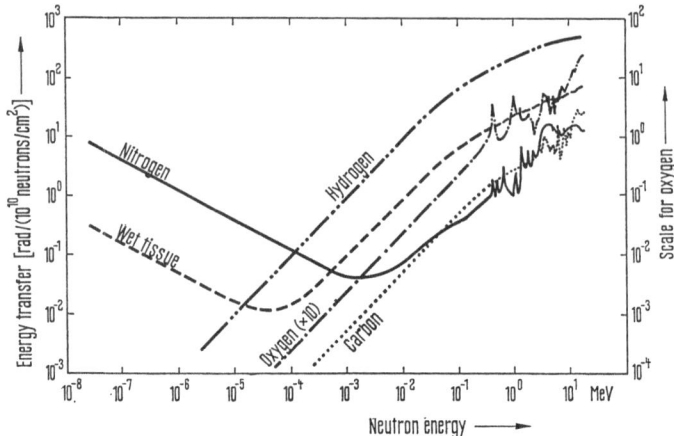

Fig. 18. Energy transfer from neutrons to hydrogen, carbon, nitrogen, oxygen, and wet tissue. (Bach and Caswell, 1968)

actions are with nuclei and not with atomic electrons is of primary significance.

a) Scattering. In materials composed of light nuclei, the most frequent interactions are elastic collisions between neutrons of moderate energy and the nuclei. The energy transferred to a nucleus may be calculated by treating the interaction as a collision between elastic spheres, based on the principles of conservation of energy and momentum:

$$E = \frac{4\,(m_n/M)}{(1+m_n/M)} \cdot E_n \cdot \cos^2 \Theta \,, \qquad (4.6)$$

where m_n and E_n represent the mass and energy respectively of the incident neutron, M and E the mass and energy respectively of the recoil nucleus, and Θ the angle between the directions of the incident neutron and the recoil nucleus. The maximum energy transfer becomes smaller as the mass of the recoiling nucleus increases, as can be seen from equation (4.6). Scattering by a hydrogen nucleus will lead, on average, to the transfer of one-half of the neutron energy, while in an interaction with a lead nucleus there is an average transfer of only about 1%. This is the reason for using hydrogeneous material (such as paraffin), rather than lead, for the screening of neutron sources.

The constituent elements of biological materials have different cross sections for neutron scattering. This is shown in Fig. 18, where the energy transfer from neutrons of different energies to hydrogen, carbon, nitrogen and oxygen, and also to wet tissue, is plotted. It can be seen that a particularly large amount of energy per incident neutron is transferred to hydrogen; this occurs firstly because it has a large neutron scattering cross

38

section, and secondly because the favourable mass ratio (equation 4.6) allows more energy to be transferred than in a collision with a heavier nucleus. As hydrogen is a major constituent of biological materials, 85 to 95% of the energy of the interacting neutrons is transferred to hydrogen nuclei (which are referred to as recoil protons), and only a minor fraction to heavier nuclei.

b) Absorption. This is another important interaction of neutrons with nuclei. A neutron is captured, producing a short-lived highly-excited nucleus (known as a compound nucleus) which in the case of light nuclei may reach a stable ground state by the emission of gamma quanta, while in the case of intermediate and heavy nuclei, protons or alpha particles may also be emitted. The probability of capture becomes smaller with increasing neutron energy, E_n, following the Fermi formula:

$$\sigma_{(n,\,\gamma)} \sim 1/\sqrt{E_n} \sim 1/v_n. \tag{4.7}$$

Table 2 shows that even for thermal neutrons, the scattering cross section is, in general, larger than the cross section for neutron capture. As the cross section for capture reactions also decreases rapidly with increasing

Table 2. *Reaction cross sections for the scattering and absorption of thermal neutrons in some biologically important elements. The abundance of the different isotopes corresponds to their natural occurrence.* (Hughes and Harvey, 1955)

Element	Scattering cross section [barn]	Absorption cross section [barn]	
H	38	0.33	(n, γ)
C	4.8	0.0032	(n, γ)
N	10	1.75	(n, p)
		0.13	(n, γ)
O	4.2	0.0002	(n, γ)

1 barn $= 10^{-24}$ cm^2

energy, practically all neutrons are slowed down to thermal or epithermal energy, by collisions within the material, before being captured. Of the elements constituting major fractions of biological materials, hydrogen and nitrogen have the largest cross-sections. The reactions induced are:

$$^1H\ (n, \gamma)\ ^2D + 2.2\ \text{MeV}\ \gamma\text{-radiation} \tag{4.8}$$

$$^{14}N\ (n, p)\ ^{14}C + 660\ \text{keV protons}. \tag{4.9}$$

However, only a small proportion of the biological action of neutrons is due to these two primary processes of collision and capture. Most of the radiation damage is caused by the action of the recoil nuclei, and by the secondary radiation produced in neutron capture processes, as shown for example in (4.8) and (4.9).

4.3. Charged Particles

The action of gamma-rays and neutrons, as already mentioned, is not a result of their primary interactions, but rather of the action of the secondary radiation, i. e. electrons and recoil protons. Consequently, the results of a study of the interactions of charged particles with matter are not only applicable to the action of fast electrons and ions, such as those produced in an accelerator, but extend also to the action of gamma-rays and neutrons. The elastic collisions of slow ions represent a special case which will be considered separately.

The electric field of a charged particle will interact with the molecular electrons in the vicinity of the path. The interaction processes can be visualized using Fig. 19, which shows a charged particle passing a molecule

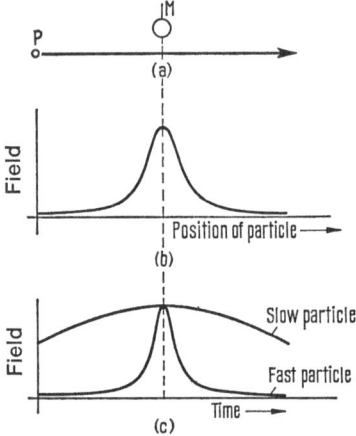

Fig. 19. Schematic representation of the interaction between a charged particle and a molecule. a The particle P passes the molecule M at a certain distance. b The magnitude of the electric field at the site of the molecule as a function of distance (velocity-independent). c The same field as a function of time (velocity-dependent)

(a). As it approaches, the electric field at the site of the molecule, and therefore the forces acting on the molecule, increase, and once the particle has passed, decrease again (b). If the electric field is plotted as a function of time (c), it can be seen that a slow particle acts on a molecule over a much longer period of time than a fast particle, thereby giving the molecule a larger momentum.

This picture, although rather simple, describes correctly some of the most important aspects of energy transfer by charged particles: 1) the interference caused by a passing particle is greater for slow particles than for fast ones; 2) the energy transfer increases with the charge of the particle; 3) the mass of the particle has no influence on the amount of energy transferred. (In the cases where the energy of the particle is specified, its velocity, of course, indirectly depends on mass.)

Quantitatively, *the differential energy loss* (or stopping power) of a charged particle, i. e. the loss of energy per unit track length, is given by the Bethe-Bloch equation:

$$-\frac{dE}{dx} = \frac{4\pi e^2 (z\,e)^2}{m\,v^2}\, n\,Z\left[\ln\frac{2\,m\,v^2}{I} - \ln(1-\beta^2) - \beta^2\right], \qquad (4.10)$$

where m is the rest mass of the electron, v and $z\,e$ the velocity and charge respectively of the particle, n the number of atoms per cm^3, Z the effective atomic number, I the mean ionization potential, and $\beta = v/c$ (c is the speed of light). The term in square brackets is strictly valid only for heavy particles; for electrons it is more complex.

The Bethe-Bloch equation (4.10) confirms the predictions 1 and 2, the loss of energy being proportional to $1/v^2$ and also proportional to $(z\,e)^2$. The third prediction, that of independence from the mass, is also confirmed. In addition, the energy loss is found to be proportional to $n\,Z$, the mean number of electrons per cm^3. Table 3 shows that this product has approximately the same value for several biologically important media.

Table 3. *Number of electrons per gram of some biologically important materials.* (Jaeger, 1959)

Material	Electrons/g
Air	$3.03 \cdot 10^{23}$
Water	$3.34 \cdot 10^{23}$
Wet tissue	$3.31 \cdot 10^{23}$
Muscle	$3.36 \cdot 10^{23}$
Bone	$3.00 \cdot 10^{23}$
Fat	$3.48 \cdot 10^{23}$
Virus protein	$3.22 \cdot 10^{23}$

The mean ionization potential I is a measure of the energy required to remove electrons from their respective states multiplied by the frequency of these events. The value of I is given approximately by the relationship:

$$I = 13.5 \cdot Z\ (e\,V)\,. \qquad (4.11)$$

Some conclusions can be drawn from a discussion of the fundamental Bethe-Bloch equation. If the expression (4.10) were accepted without reservation, then a catastrophe would be expected at low particle speeds, since the energy transfer should become infinitely large. This is a result of the assumption that the charge of the particle ($z\,e$) remains constant, which is not valid under these conditions. For example, an α-particle passing through matter is not always doubly charged, as it may capture an electron and continue its flight as a singly charged, and therefore less strongly inter-

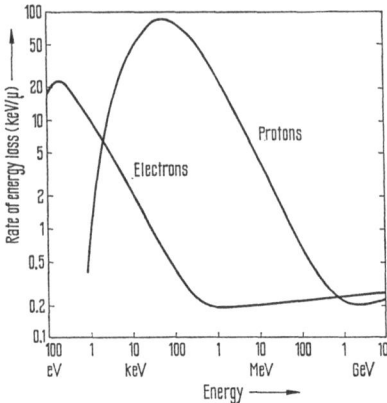

Fig. 20. Differential energy loss dE/dx (stopping power) of electrons and protons in water, as a function of energy. (Lewis, 1954; Neufeld and Snyder, 1961)

acting, helium ion. The capture probability increases as the particle slows down. At a sufficiently low speed, the singly-charged helium ion picks up another electron, thereby becoming an helium atom of even lower ionization density. These processes can be taken into consideration by introducing a velocity-dependent particle charge into the Bethe-Bloch formula. Barkas (1963) suggested the following expression:

$$z^* = z \left[1 - \exp\left(-125 \, \beta \, z^{-2/3}\right)\right]. \tag{4.12}$$

At low speeds $(\beta = v/c \to 0)$, z^* tends to 0 in such a way that the whole expression dE/dx also approaches 0. On the other hand, as dE/dx also falls off as $1/v^2$ at higher energies, the energy loss must pass through a maximum at low particle energies. This maximum, known as the "Bragg Peak", can be seen in Fig. 20, in which the stopping power for electrons and protons is shown as a function of their energy. For electrons the maximum occurs at approximately 200 eV, and for protons between 60 and 100 keV. The slight rise at very high energies is a relativistic phenomenon described by the last two terms of the Bethe-Bloch equation.

The Bragg maximum can be studied experimentally by inserting absorbers of varying thickness into the particle path, and then measuring the differential ionization density using a thin ionization chamber. With increasing absorber thickness, the energy of the particles is reduced, whereupon the energy transfer increases. However, once past the "Bragg Peak" it falls off rapidly to zero (see Fig. 21).

Linear Energy Transfer (LET). Because of its great importance in the specification of radiation quality, the differential coefficient dE/dx has been given a special name. It is known as the "linear energy transfer" (LET), which in mathematical expressions will be referred to as L. It is measured in keV per μ. Dividing this by the density ϱ of the material, the quantity

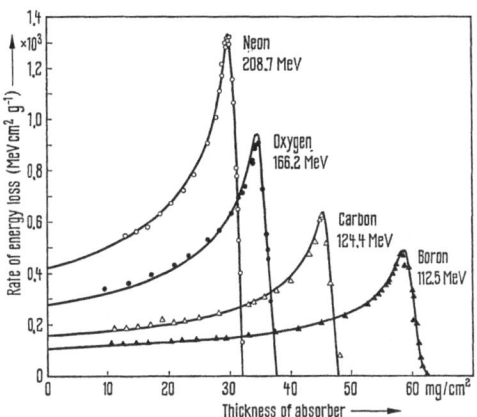

Fig. 21. Bragg curves of some heavy ions in tissue-equivalent material. (Brustad, 1961)

L/ϱ which is independent of density is obtained, and this is also generally referred to as LET, although sometimes called the mass stopping power. This quantity is commonly measured in $MeV \cdot cm^2 \cdot g^{-1}$. When $\varrho = 1$ this quantity is larger than the LET measured in keV/μ by a factor of 10. Table 4 shows the LET values for several different radiations. The lowest possible rate of energy transfer is $0.2\ keV/\mu$; this fact is clearly shown in Fig. 20. The LET for different kinds of radiation may vary by several orders of magnitude (Table 4). It is to be expected that radiations with such different

Table 4. *Compilation of LET values for various types of radiation* ($\varrho = 1\ g/cm^3$)

Type of radiation	LET [keV/μ]
8 MeV γ-rays	0.2
^{60}Co γ-rays	0.3
200 keV X-rays	2.5
340 MeV protons	0.3
2 MeV protons	17
27 MeV α-particles	25
5 MeV α-particles	90
3.4 MeV α-particles	130
100 MeV carbon ions	160
160 MeV neon ions	450
330 MeV argon ions	1300

physical properties will also differ in their biological effectiveness. This point will be discussed in greater detail in Chapter 5.

Range of Charged Particles. The ionization density of a particle decreases rapidly after it has passed through a certain thickness of absorber,

43

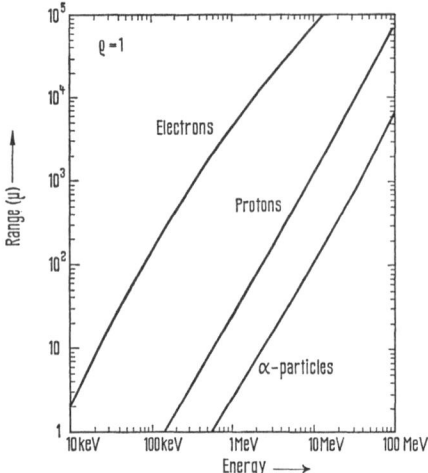

Fig. 22. Range of electrons, protons, and alpha particles in organic materials of density $\varrho = 1$ g/cm³, as a function of energy. (Bacq and Alexander, 1961)

and reaches zero when the particle comes to rest (Fig. 21). This range is quite well defined for heavy charged particles. It depends on the charge and energy of the particles, and can be evaluated for medium and high energies by integration of the Bethe-Bloch formula. The range-energy curves for electrons, protons and alpha-particles in water are shown in Fig. 22. As biological materials contain approximately the same number of electrons per gram as water (see Table 3), the same curves are valid for nucleoprotein and tissue; however, allowance must be made for the different densities of the materials. At lower energies the theory becomes inaccurate and therefore the values calculated for the range are unreliable. Under these circumstances it is necessary to rely on measurements with foils, although there are some experimental difficulties at low energies. Measurements in gases have the disadvantage that the results cannot be applied to condensed matter without certain reservations. There have also been some attempts to measure the ranges of slow electrons and protons using thin layers of enzymes (Davis, 1954; Person *et al.*, 1963).

Elastic Nuclear Collisions. Although the interaction with electrons is the most important mechanism by which energy is transferred from charged particles, there is a corresponding interaction process with the nucleus, known as elastic nuclear collision. Similar considerations can be carried out for this process as were carried for electron interactions. The Coulomb field of the charged particle interacts with the Coulomb field of a nucleus. If a sufficiently high momentum is imparted, then the atom concerned can be removed from its molecule. As much larger masses have to be moved in nuclear collisions than in the case of ionization, the incident particle must be moving sufficiently slowly and must pass very close to the nucleus. This is the reason why the probability of nuclear collisions at high energies is smaller by about three orders of magnitude than that of interactions with electrons. As the cross section for elastic nuclear collisions increases with decreasing particle energy, even below the "Bragg Peak" (i. e. the region of low particle velocities where the probability of ionization is greatly reduced), at low energies more energy is transferred by nuclear collisions than by interactions with electrons. With protons this occurs below an energy of 1.5 keV (Neufeld and Snyder, 1961). More details of this primary process are described in the review article by Jung and Zimmer (1966).

4.4. Uptake of Energy by Molecules

The energy transfer processes will now be considered from the point of view of the atoms and molecules, by enquiring what "energy packets" are imparted to a molecule by a passing charged particle. This question is far more relevant to the induction of biological changes than the fate of the ionizing particle discussed so far. The Bethe theory contributes relatively little to the answer, as it is based on an ingenious use of theoretical "sum rules", thereby avoiding the description of where the energy has actually gone (cf. Platzman, 1967).

Oscillator Strength. A realistic discussion of energy absorption at the molecular level is made possible by the introduction of the term "oscillator strength". In the following, the terminology chosen by Platzman (1962) will be adopted. The simplest case, that of the hydrogen atom, is considered first. An atomic electron can be transferred to discrete higher energy states by the absorption of energy; in the subsequent transition to the ground state this energy is released again as electromagnetic radiation of a specific wavelength. Even in a small molecule, the situation becomes much more complex, since as well as the excitation of electrons to higher energy states, oscillations and rotations of individual groups within the molecule can also occur. To enable such a complex structure to be treated mathematically, the excitation of the oscillator s (of energy $E_s = h\nu_s$) is referred to, rather than the excitation of specific electronic, oscillatory, and rotatory levels. The different oscillators are not adequately specified by their energy levels; in addition, their effective numbers representing the probabilities of their excitation must be known. This frequency is called the *oscillator strength* "f_s" of the oscillator having the frequency ν_s. These numbers are normalized by making their sum equal to the total number of electrons in the molecule:

$$\sum_s f_s = Z . \tag{4.13}$$

From the definition of oscillator strength it can be seen that the macroscopic optical absorption coefficient μ is proportional to the oscillator strength

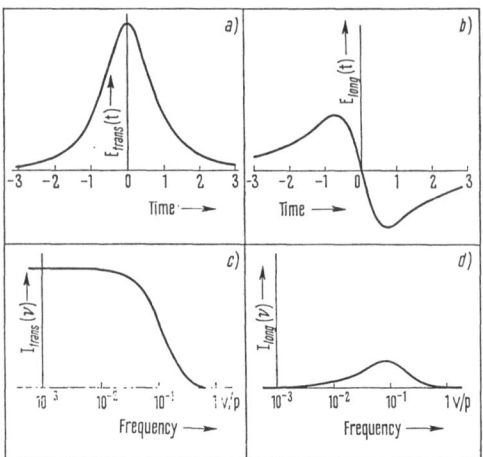

Fig. 23. Action of a fast charged particle on a molecule (velocity: v; impact parameter: p). a and b The time-dependent changes in the electric field parallel to the direction of motion of the particle (longitudinal component) and perpendicular to it (transverse component). c and d Spectrum of virtual photons, resulting from a Fourier analysis of the field in a and b. $I(\nu)$ ist the spectral intensity of the photons passing through the unit area at the position of the molecule, during the passage of the particle. (Platzman, 1962)

at the corresponding frequency: $\mu_s \sim f_s$. For transitions within a continuous spectrum, $df/d\nu$ takes the place of the discrete oscillator strengths and therefore:

$$\mu(\nu) \sim df/d\nu . \qquad (4.14)$$

If a molecule were exposed to a light source delivering equal numbers of photons in each frequency interval from visible light up to X-rays ("white light"), then the number of molecules activated to the state s would be proportional to the oscillator strength of this state:

$$N_s = \text{const} \cdot f_s . \qquad (4.15)$$

Energy Transferred by Charged Particles. It is now necessary to investigate which oscillators are activated by a fast charged particle passing within a given distance of a molecule. The force acting on the molecular electrons (see Fig. 19) can be split into two components: one parallel to the path of the particle, the longitudinal component, and the other perpendicular to it, the transverse component (Fig. 23). Each of these components can be represented by a Fourier series, as a sum of purely harmonic functions of time:

$$E_{\text{trans}} = \Sigma \, k_i \cdot \cos (2\,\pi\,\nu_i\,t) . \qquad (4.16)$$

The longitudinal component is so small that it can be neglected. The intensity of the transverse component, in contrast, is constant from low frequencies up to almost the maximum frequency, i. e. each frequency interval includes the same amount of energy of these "virtual photons":

$$I(\nu) = \text{const} . \qquad (4.17)$$

The number of photons per frequency interval is obtained by dividing by the photon energy:

$$n(\nu) = \text{const}/h\nu . \qquad (4.18)$$

This means *that the action of a charged particle passing through matter corresponds to exposure to "white light" with a frequency distribution proportional to $1/h\nu$*. Equation (4.15) showed that with a constant frequency distribution, N_s is proportional to the oscillator strength. If the frequency of the incident light has a $1/h\nu_s = 1/E_s$ distribution, then the number of molecules excited to the state s by the passage of a fast charged particle is given by:

$$N_s \sim f_s/E_s . \qquad (4.19)$$

The analogous expression obtained for continuous spectra is:

$$N(E) \sim (df/dE)/E . \qquad (4.20)$$

This equation is called the "optical approximation". Its scope and limitations have been discussed by Platzman (1967).

The oscillator strength f_s is approximately proportional to the number of electrons in the shell in which the activation is initiated, i. e. it becomes larger for the outer shells. E_s is approximately proportional to the square of the effective nuclear charge acting on this transition, and is therefore smaller for the outer shells because the inner electrons shield the nuclear field. This means that *the predominant primary process resulting from the passage of a charged particle through matter is the excitation of valence electrons*.

Oscillator strengths of multi-atomic molecules. A rigorous quantum-mechanical calculation of oscillator strength has as yet been carried out only for atomic hydrogen. A reasonably accurate picture of the spectrum of oscillator strengths of simple molecules can be obtained by taking into consideration various experimental observations, such as optical absorption, inelastic scattering of electrons, light scattering, dielectric constants, etc. However, only the spectra for relatively simple molecules in the gaseous phase can be obtained in this way. Fig. 24 shows the excitation spectrum $R \cdot (df/dE)/E$ (cf. equation 4.20) of methane constructed using these methods (Platzman, 1962). The Rydberg constant R is introduced to obtain dimensional balance. The results show that excitations are most frequent at relatively high energies and that an important part of the excitation spectrum is above the ionization potential. However, not all states with energies above the ionization potential necessarily lead to ionization of the molecule. These "superexcited states" may dissipate their energy by intramolecular changes or by splitting the molecule into two radicals (dissociation), instead

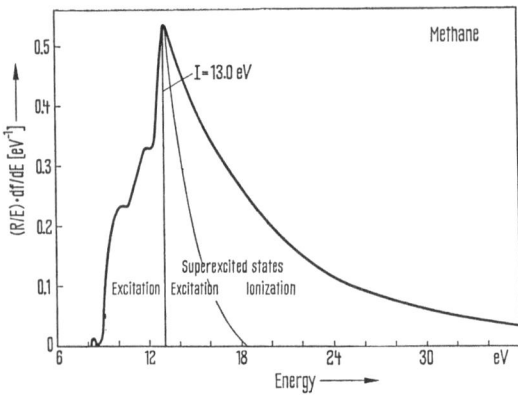

Fig. 24. Excitation spectrum of methane. (Platzman, 1962)

of ionization. The part of the spectrum marked "ionization" refers to those transitions which always lead to the loss of an electron.

It would be extremely useful to know the excitation spectrum of biologically important macromolecules, as it could be used to determine the distribution and frequency of the transferred energy packets. A theoretical calculation is not possible, and the experimental approach is so complicated that nobody has attempted it as yet. However, some results will be presented in the last section of this chapter which give some qualitative indications of the excitation spectra of biological molecules. In general, the following can be said: for most organic molecules the majority of oscillator strengths lie approximately 10 to 30 eV above the ground state. All of these molecules also have oscillators at longer wavelengths, although their strength is very small. If the molecule has multiple-bonds, there is a greater probability of exciting the low-lying states, which is reflected by the increased absorption of near-ultraviolet or visible light. This tendency is even more pronounced in the presence of conjugated double-bonds. However, even in these special cases most of the spectrum of oscillator strengths is above the ionization potential, i. e. the energy required to remove the least tightly bound electron is usually well below the main band of oscillator strengths occurring at 20 eV or higher.

The excitation spectrum is essentially independent of the type and velocity of the particle. Consequently the same states are always excited with the same frequency, so that the magnitude and the relative frequency of the transferred energy is not influenced by the quality of the radiation. This means *that at the molecular level, all kinds of ionizing radiations have qualitatively the same effect.*

An exception to this rule is observed when UV light of very short wavelength, so-called vacuum-ultraviolet, is used. This technique has not yet been extensively used for radiation biological experiments because of the experimental difficulties it presents. It allows specific amounts of energy to be transferred to the irradiated material, i. e. the excitation of specific levels, which may be well above the ionization potential. More will be said about these experiments in Chapter 6.

4.5. The Energy Distribution of Secondary Electrons

With the information acquired from the previous sections, the energy distribution of the secondary electrons generated by charged particles can now be discussed. It is useful to consider two different mechanisms by which a charged particle can release secondary electrons from atoms or molecules: knock-on and glancing collisions.

Knock-on Collisions. These interactions are described fairly accurately by classical electrodynamics. They occur when the speed of the incident particle is many times greater than the orbital speed of the atomic electrons, and when the particle comes into close proximity with an electron. The time of interaction in this case is so short that the atomic electron can be considered to be stationary. The transferred momentum is generally so large that the binding energy of the electron can also be neglected. The scattering of the emitted electrons follows Rutherford's equation, and their energy distribution decreases proportionally with $1/E^2$. The probability of knock-on collisions depends only on electron density, and not on the chemical composition of the irradiated material.

Glancing Collisions. In contrast, the frequency of the glancing collisions depends on the oscillator strength, and therefore on the chemical composition of the material. In this process, the charged particle need not pass very near to the atom, as it can interact via its electrostatic field with atoms and molecules at a relatively greater distance. The distance, or impact parameter, can be as large as 1000 Å or more in gases (Fano, 1954), while in condensed matter collective excitations occur extending over a volume of (100 Å)³ or more (Fano, 1961). The orbital motion of atoms cannot be neglected in glancing collisions, in contrast to knock-on collisions. In addition, the molecules may become polarized, which will in turn affect the strength of the electron bonds. For these reasons glancing collisions cannot be adequately described by simple electrostatics, and the corresponding theoretical calculations for larger atoms or molecules become extremely difficult. Glancing collisions occur, in general, 8 to 10 times more frequently than knock-on collisions (Fano, 1952).

The *energy distribution of the secondary electrons* produced via these two mechanisms by the action of 100 keV electrons, will now be considered. Because of the difficulties mentioned above, the results shown in Fig. 25 have been calculated for hydrogen atoms. This example can only be used for purely qualitative considerations, since hydrogen atoms obviously bear little resemblance to the complex macromolecules found in biological systems.

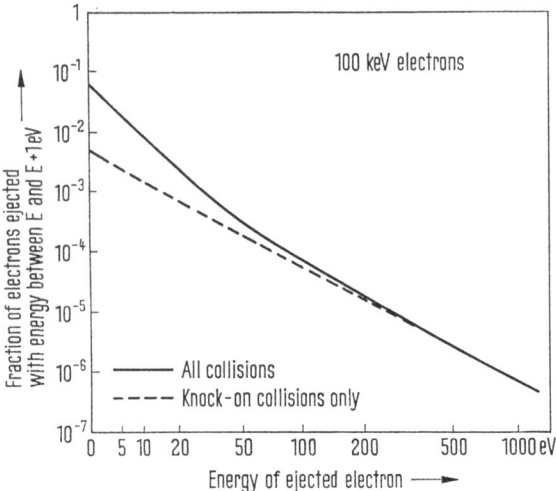

Fig. 25. Energy distribution of the secondary electrons released by 100 keV electrons, calculated for glancing collisions and knock-on collisions with hydrogen atoms. (Lewis, 1954)

Secondary electrons with "zero" kinetic energy occur most frequently, and the number of secondary electrons falls off rapidly with increasing energy. At low energies this decrease is approximately proportional to $1/E^3$ and at higher energies to $1/E^2$. Most of the low energy electrons originate from highly excited molecular states (glancing collision), and only about 1 in 10 from a knock-on collision between an incident particle and an atomic electron. As would be expected, the relative contribution of "knock-on" ionizations increases with increasing kinetic energies of these released electrons.

A more useful picture of the relative contribution of secondary electrons with low and moderate energy is obtained if their total energy (expressed as a fraction of the total energy of the primary particles) is examined, rather than their numbers (Fig. 26). As an electron slows down from 500 keV to 400 keV in water, about 17% of the total transferred energy appears as excitations of the water molecules, while 35% is transferred to secondary electrons with kinetic energies below 100 keV. The energy converted to X-rays is negligible, but the emission of K-electrons from oxygen accounts for 19% of the energy; however, the K-ionization process requires a relatively large amount of energy, and the number of molecules damaged by this primary process is negligible compared with the number damaged through the ionization of valence electrons.

Primary Ionization. About 70% of the energy of the primary particles is transferred to secondary electrons that have sufficient energy to cause further ionizations (Fig. 26 non-shaded area). Ionizations caused by the lower-energy secondary electrons of this group will occur in the immediate

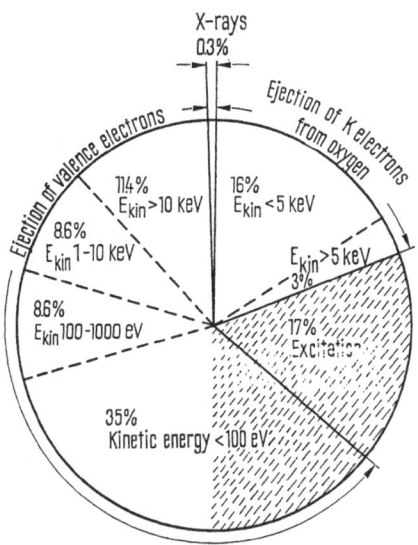

Fig. 26. Partition of the energy deposited by the slowing-down of electrons from 500 keV to 400 keV in water. The unshaded area represents the fraction of the energy transferred to secondary electrons which then have sufficient energy to cause further ionization. (Lewis, 1954)

vicinity of the primary ionization event. Thus, a number of ion pairs (positive ion plus electron), all of which originate from the same primary ionization by the incident particle, are produced within a relatively small volume, which is referred to as an "ion cluster". As high energy secondary electrons are comparatively rare, the probability of the occurrence of clusters consisting of many ion pairs decreases rapidly (Table 5). In the following, primary ionization, ion cluster, and energy loss event are used as equivalent expressions.

Table 5. *Frequency distribution of ion pairs per ion cluster as determined from cloud chamber pictures.* (Ore and Larsen, 1964)

Ion pairs	Frequency ($^0/_0$)
1	63.3
2	20.4
3	9.2
4	4.1
5	2.0
6	1.0

Delta Rays. Secondary electrons with much higher energies also occur, as can be seen from Fig. 26. In cloud-chamber pictures, these energetic electrons appear not as clusters but, owing to their greater range, as short tracks branching from the track of the primary particle. These secondary

51

electrons are called δ-rays if their energy exceeds 100 eV (Lea, 1946; Pollard *et al.*, 1955), although this limit is sometimes put at 1 keV (Bacq and Alexander, 1961); this is, however, merely a matter of convention. In theory, the total primary energy of an electron can be transferred to a δ-ray. The maximum energy that can be transferred by heavy primary particles of mass M is limited to $4\,m_e/M$, according to the principles of conservation of energy and momentum. For example, a 1 MeV proton can transfer a maximum of 2 keV to an electron. The intensity and energy distributions of the secondary electrons released by different radiations have been compiled in numerous tables by Lea (1946). In unfavourable circumstances, a large proportion of the energy of the primary particle may be carried, by an energetic delta-ray, out of the biological structure in which the primary absorption event has occurred; this leads to various complications in calculations of the size of the biological targets (see Chapter 5).

Irradiation with fast electrons (LET $= 0.2$ keV/μ) invariably produces secondary electrons with widely varying energies, including those between 100 and 500 eV whose energy transfer is two orders of magnitude greater than that of the primary particle (see Fig. 20). On the other hand, a considerable proportion of the energy of densely ionizing primary particles is deposited by δ-rays, so that in general no specific value for the rates of energy transfer can be given for the irradiation of biological objects. This point also demonstrates some of the problems inherent in the concept of linear energy transfer (LET), introduced in the previous section. These problems are demonstrated particularly clearly by a comparison of the range of the primary particle with the sum of the ranges of the released secondary electrons. For 400 keV electrons, the total range of the delta-rays represents less than 3% of the range of the primary electron, whilst a 1 MeV alpha-particle has a range less than one-half of the total range of the delta-rays (Lea, 1946).

4.6. Energy Deposited per Primary Interaction

The amount of energy which *on average* is transferred to matter by one primary ionization process is of special interest for radiation biology, and in particular for the target theory. This quantity has experienced a dramatic "devaluation" over the years. From the analysis of cloud-chamber pictures, Pollard and colleagues (1955) arrived at the value of 110 eV per ion cluster. More recent evaluations of the number of ion pairs per cluster (Ore and Larsen, 1964; see Table 5) enabled the energy expenditure per primary event to be calculated as 54 eV, with 33 eV being deposited per ion pair. However, there are justifiable doubts as to whether results obtained in gases can be extrapolated to condensed matter, since the occurrence of collective excitations in a solid leads to a distribution of oscillator strengths different from that in a gas (Fano, 1961).

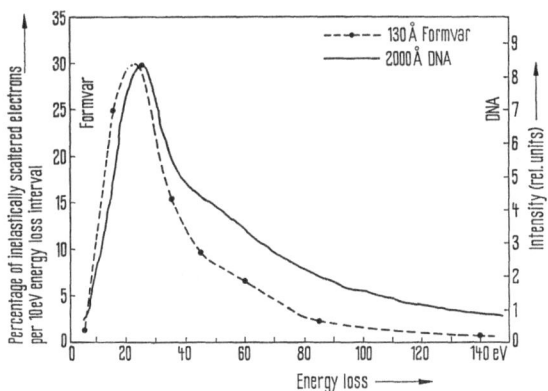

Fig. 27. Distribution of energy loss events of electrons passing through thin layers of organic material. $--\bullet--$ passage of 20 keV electrons through a Formvar foil of thickness 130 Å. (Rauth and Simpson, 1964); ——— passage of 150 keV electrons through a 2 000 Å thick DNA film; scattering angle: 51.5″. (Johnson and Rymer, 1967)

Direct measurement of the energy loss per primary ionization in condensed matter is, therefore, particularly important. Rauth and Simpson (1964) passed 20 keV electrons through thin Formvar foils, and determined the frequency of the various energy loss events from the amount of inelastic scattering that occurred. For example, the probability that there will be more than one energy loss event in the passage of one particle through a foil 130 Å thick is very small, and the results shown in Fig. 27 (broken line) can, therefore, be considered to be a measure of the frequency distribution of the different energy loss events. It is remarkable that only rarely less than 10 eV is transferred per primary interaction. The amount of energy most frequently transferred is 22 eV, while 60 eV is the average per energy loss event. The energy distribution can be compared with the numbers of ion pairs occurring in the clusters in a gas, and this comparison will be useful in the discussion of the track segment method (Chapter 5.3). Although the energy distribution (Fig. 27) gives no direct information about the number of ion pairs, these two quantities are complementary and can be related by attributing a certain mean energy for the formation of an ion pair. In one way, this is equivalent to saying either that a particular kind of molecular damage requires, for example, two primary events each consisting of two ion pairs, or alternatively that a corresponding specified amount of energy is involved. In either case, the probability of the occurrence of such an event decreases with increasing energy or number of ions required per primary event (Fig. 27 and Table 5).

Fig. 27 shows that the spectra of energy loss events in DNA and Formvar are fairly similar. However, since it is difficult to obtain DNA in layers as thin as Formvar, more than one energy loss event per passing particle may occur occasionally. This is reflected in the slower falling-off of the spectrum above 40 eV. If correction is made for these multiple events, and in addition the contribution from knock-on collisions is subtracted, then this curve gives an approximate picture of the excitation spectra (oscillator strength divided by the energy of the corresponding state) of DNA. The realization of this is so difficult (see the discussion in Chapter 4.3), that it is not possible to go further than this qualitative statement.

The primary processes have been considered from the point of view of the interacting particles as well as from that of the damaged molecules. The first approach was found to be more accessible than the second, both experimentally and theoretically. However, the second aspect certainly has a much greater significance in the study of radiation biology. The further developments of radiation physics, therefore, seem likely to be of little importance for the understanding of the biological action of radiation, whereas additional information concerning the transfer of energy to the irradiated molecule, so far rather sparse, will be important for the development of radiation biology.

References

Bach, R. L., Caswell, R. S.: Radiat. Res. **35**, 1 (1968).

Bacq, Z. M., Alexander, P.: Fundamentals of radiobiology. New York: Pergamon Press 1961.

Barkas, H.: Nuclear research emulsions, Vol. 1. New York: Academic Press 1963.

Brustad, T.: Radiat. Res. **15**, 139 (1961).

Davis, M.: Phys. Rev. **94**, 243 (1954).

Fano, U.: In: Symposium on radiobiology. Ed.: J. J. Nickson. New York: John Wiley & Sons 1952, p. 13.

— In: Radiation biology I, 1. Ed.: A. Hollaender. New York: McGraw-Hill 1954, p. 1.

— In: Comparative effects of radiation. Eds.: M. Burton, J. S. Kirby-Smith, and J. L. Magee. New York: John Wiley & Sons 1961, p. 14.

Hughes, D. J., Harvey, J. A.: Brookhaven nat. Lab. Rept. 325 (1955).

Jaeger, R. G.: Dosimetrie und Strahlenschutz. Stuttgart: Thieme 1959.

Johnson, C. D., Rymer, T. B.: Nature **213**, 1045 (1967).

Jung, H., Zimmer, K. G.: In: Current topics in radiation research, Vol. II. Eds.: M. Ebert and A. Howard. Amsterdam: North-Holland Publ. Co. 1966, p. 69.

Lea, D. E.: Actions of radiations on living cells. Cambridge: University Press 1946.

Lewis, M.: Cited by Fano (1954).

Neufeld, J., Snyder, W. S.: In: Selected topics in radiation dosimetry. Vienna: Internat. Atomic Energy Agency 1961, p. 35.

Ore, A., Larsen, A.: Radiat. Res. **21**, 331 (1964).

Person, S., Hutchinson, F., Marvin, D.: Radiat. Res. **18**, 397 (1963).

Platzman, R. L.: Vortex **23**, 372 (1962).

— In: Radiation Research. Ed.: G. Silini. Amsterdam: North-Holland Publ. Co. 1967, p. 20.

Pollard, E. C., Guild, W. R., Hutchinson, F., Setlow, R. B.: Progr. Biophys. **5**, 72 (1955).

Rauth, A. M., Simpson, J. A.: Radiat. Res. **22**, 643 (1964).

Spencer, L. V., Stinson, F.: Cited by Fano (1954).

White, G. R.: Cited by Fano (1954).

Chapter 5. Target Theory and Action Cross Section

Now that the most important primary processes of energy absorption have been considered (Chapter 4), and powerful formalisms for the description of dose-response relationships have been developed (Chapter 2 and 3), the unavoidable problem arises of linking the formal description with physical reality. The resultant systematic presentation of this chapter is therefore particularly valuable, as its conclusions will provide some important aids for the better understanding and assessment of many radiation biological phenomena (for example, the temperature and oxygen effects). The considerations of this chapter require first of all a precise definition of the term "hit", as this was not possible in the previous discussion of the hit theory.

5.1. Establishment of a Rigid Concept of a "Hit"

The basic idea of the target theory can be expressed as an attempt, based on physical considerations, to determine the number of hits per unit volume, and to derive from this the volume of the target and, therefore, the size of the radiation-sensitive substructures of the biological system. The discussion of the hit theory has already shown that this is in general a difficult, if not impossible undertaking. The discussion at this stage will, therefore, be limited to single-hit processes. In practice, this is not a serious limitation since, with the exception of some examples of irradiation in dilute aqueous solution, there is scarcely a situation in the field of molecular radiation biology where a dose-response curve with a shoulder has to be interpreted as a multi-hit curve; in any case, exponential dose-effect curves are usually obtained, especially when corpuscular radiation is used. These observations indicate that in the case of ionizing radiation, a hit depends on a single transfer of a certain amount of energy, the magnitude of which depends on the kind of damage and the system irradiated. It could, therefore, be said that the test effect occurs when a certain minimum amount of energy is transferred by the passage of an ionizing particle (e. g. a secondary electron) through a sensitive region. This sounds both obvious and simple. However, the determination of the target size will depend on the probability that this minimum energy is transferred. The calculation of this probability based on the energy distribution as, for example, measured by

Rauth and Simpson (1964) (Fig. 27), represents a complex mathematical problem. In contrast, the absorption of a UV-quantum can be considered to represent a hit. This in itself does not, however, verify the concept of a target, since the biologically important absorptions are resonance processes, i. e. they depend to a large extent on the wavelength.

5.2. Target Theory

A procedure for determining the size of the target, which seems an obvious application of the basic hit theory, will be outlined before continuing with a rigorous treatment of the target theory. Following the terminology previously used, the calculation will be carried out in units of mean energy per primary ionization. This, according to the measurements of Rauth and Simpson (1964), is 60 eV (see Fig. 27). In the following greatly simplified calculations, a hit is assumed to be equivalent to a mean energy absorption of 60 eV. It is known from hit theory that the nominal target volume v for single-hit curves is equal to the reciprocal of D_{37}, i. e. proportional to radiation sensitivity. The calculation involves the conversion of the dose unit "rad" used nowadays into the unit hits per cm³ or per g used in the hit theory, to enable the target size to be determined. The "rad" is defined by:

$$1 \text{ rad} = 100 \text{ ergs/g} = 6.24 \cdot 10^{13} \text{ eV/g} . \tag{5.1}$$

If 60 eV is taken as the mean hit energy, this leads to:

$$1 \text{ rad} = \frac{6.24 \cdot 10^{13}}{60} = 1.04 \cdot 10^{12} \text{ hits/g} . \tag{5.2}$$

The mass of the target is, therefore, given by:

$$M = \frac{1}{D_{37}} \frac{1}{1.04 \cdot 10^{12}} = 0.96 \cdot 10^{-12}/D_{37} \tag{5.3}$$

where M is measured in grams and D_{37} in rads. The target is obtained by dividing by the density ϱ:

$$v = \frac{M}{\varrho} = \frac{0.96 \cdot 10^{-12}}{\varrho \cdot D_{37}} \text{ [cm}^3\text{]} . \tag{5.4}$$

Finally, the multiplication by Avagadro's number $6.022 \cdot 10^{23}$ gives the molecular weight of the target:

$$MW_T = 5.8 \cdot 10^{11}/D_{37} \text{ [Dalton]} . \tag{5.5}$$

Using this equation, "target molecular weights" (MW_T) of numerous enzymes have been calculated from their D_{37}, and compared with the molecular weights determined by physico-chemical methods. A straight line plotted at 45° (in Fig. 28) is to be expected if the two molecular weight determinations give identical results. Although in many cases the deviation of the results from this straight line is larger than the experimental error,

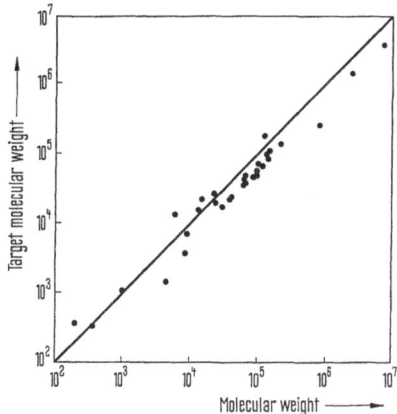

Fig. 28. Comparison of molecular weights of various enzymes with the molecular weight of the target as calculated from the radiation sensitivity, using equation (5.5). (Pollard, 1959)

the agreement shown in Fig. 28 is surprisingly good over several orders of magnitude. "Surprisingly", because this correlation implies that the sensitivity of a molecule increases with its size. At first sight, it might be concluded that an energy transfer of 50 to 100 eV will always cause inactivation. A blemish on this interpretation of these experimental observations (Fig. 28) is that the best agreement is found at room temperature, even though there is no apparent reason for this. At low temperatures, where the misleading indirect effects are almost negligible (see Chapter 7), smaller values are generally obtained. This example, together with numerous cases where the target molecular weight MW_T is unrealistic, show that the hopes that ionizing radiations could be used to determine the molecular weight of macromolecules unobtainable in a pure form cannot be fulfilled.

5.3. Theory of the Action Cross Section

A more rigorous and general treatment of the target theory will now be undertaken. The definition of a hit given in Chapter 5.1 will be retained: the fact that a minimal and not, as assumed in the previous section, a mean energy transfer to the system is required will be taken into account. It will become apparent that this makes the target theory an alternative concept to the hit theory. While according to the hit theory the test effect occurs once the biological system has received a minimal number of hits, it is now assumed to occur when an ionizing particle deposits a certain minimal amount of energy in a single energy loss event. If this does not happen, then no response is to occur. In the target theory, it is therefore not necessary to assume an accumulation of "prior" energy loss events, so that when there is only a single target (which will be assumed in the following) an exponential dose-response curve is always obtained. Not all energy loss events will be effective, as there is a limiting minimal amount of energy required for a test effect, and this is taken into consideration by introducing the "action cross section". This quantity is obviously LET-dependent, since the probability that a given amount of energy will be deposited per unit length of the particle track increases with increasing linear energy transfer.

These considerations, and in particular the use of the term LET, refer primarily to corpuscular radiation, i. e. energetic ions or electrons. Consequently the occurrence of a hit in the target v can be correlated with the passage of a charged particle through a formally defined cross section, which will be referred to as the action cross section σ. An incident parallel beam of corpuscular radiation, is described by the term particle fluence, F, rather than the dose, i. e. the total number of particles which have passed through the unit area after a dose D. A single-hit curve can, therefore, be represented in two different ways:

$$N/N_0 = e^{-vD}; \quad \text{where} \quad v = 1/D_{37}$$
$$N/N_0 = e^{-\sigma F}; \quad \text{where} \quad \sigma = 1/F_{37} \tag{5.6}$$

where dose and fluence are related by:

$$D = F \cdot L/\varrho \tag{5.7}$$

(L referring to the linear energy transfer).

From this, the following relationship between the target ($v = 1/D_{37}$) and action cross section is obtained:

$$1/D_{37} = \frac{\varrho}{L} \cdot \sigma(L). \tag{5.8}$$

This equation links the target theory with the theory of the action cross section, which will now be developed. The dependence of the action cross section σ on LET can be visualized qualitatively as follows: the amount of energy necessary to constitute a hit is only occasionally transferred during the passage of a sparsely ionizing particle. However, the probability increases with LET, until eventually, the passage of every particle initiates the test effect. According to these "classical" ideas, the action cross section would be expected to approach the geometrical cross section of the irradiated objects, or alternatively that of a particularly sensitive substructure, as LET increases. That these ideas are basically correct, i. e. that the action cross section in fact behaves like a geometrical cross section, is well illustrated by the following experiment. The rod-shaped tobacco mosaic virus (TMV) can be orientated by a suitable procedure. If such orientated preparations are irradiated with a parallel beam of 4 MeV deuterons, and the inactivation cross section is plotted against the orientation of the rods relative to the direction of these particles, then results such as those shown

58

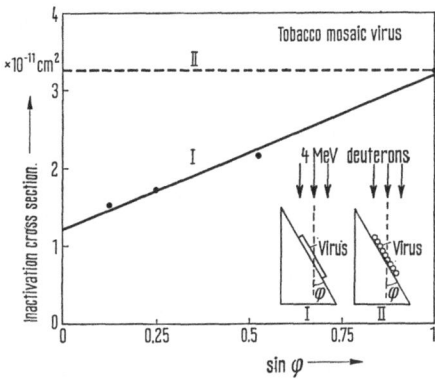

Fig. 29. Cross section for the inactivation of orientated preparations of tobacco mosaic virus with 4 MeV deuterons as a function of their direction relative to the beam. I: The rotational angle φ is equal to the angle between the direction of the beam and the axis of the virus particles. II: The virus axis is perpendicular to the direction of the beam. (Pollard and Whitmore, 1955)

in Fig. 29 are obtained. When the rotary angle φ is equal to the angle between the incident particles and the orientation of the virus (Case I) then the inactivation probability is proportional to sin φ. If, however, the virus is rotated in such a way that the direction of the radiation is always perpendicular to the orientation of the virus (Case II), the inactivation cross section remains constant, as would be expected, and is equal to the maximal cross section of Case I.

a) The Track Segment Method

The dependence of the action cross section on LET can now be calculated on the basis of the above considerations, using the "track segment method" of Howard-Flanders (1958). In order to express the convergence of the action cross section to the geometrical cross section σ_g, it will be written as follows:

$$\sigma(L) = \psi(L) \cdot \sigma_g \; ; \; \text{where} \; \psi(0) = 0 \; ; \; \psi(\infty) = 1 \, , \tag{5.9}$$

where $\psi(L)$ is the probability that a particle of LET L passing through the target will initiate the test effect, and is therefore referred to as the action probability. The rigorous calculation of the functions $\sigma(L)$ and $\psi(L)$ is fairly complex. This is partially due to the definition of a hit, in which the distribution of energies must be considered as well as the discreteness of energy loss events, (see Fig. 27). The discussion will, therefore, be limited to the case in which the hit energy $E_{\min}$ is, as a first approximation, set equal to an integer multiple of a mean energy loss event $\overline{E}$:

$$E_{\min} = n \cdot \overline{E} \, . \tag{5.10}$$

59

$\overline{E}$ can, for a given LET, be expressed by the average number $\overline{y}$ of energy loss events per particle:

$$\overline{E} = \frac{L \cdot d}{\overline{y}}, \qquad (5.11)$$

where d has the dimension of a length and is called the "track segment", following Howard-Flanders (1958). It will generally be comparable with the average diameter of the geometrical target. The frequency of the energy loss $\overline{E}$ is assumed to follow a Poisson distribution, i. e. the probability of exactly y such events occurring is given by:

$$P(y) = \frac{\overline{y}^y e^{-\overline{y}}}{y!}. \qquad (5.12)$$

On the basis of considerations analogous to those used in the derivation of multi-hit curves, the probability that at least n energy loss events will have occurred per particle passage, i. e. the action probability, is given by:

$$\psi(L) = 1 - e^{-\overline{y}} \sum_{k=0}^{n-1} \frac{\overline{y}^k}{k!}. \qquad (5.13)$$

The equations (5.10), (5.11) and (5.13) represent only a first approximation to the problem, as already mentioned, since no allowance has been made for the fact that the energy loss events occur in the form of ion clusters. These in turn consist of one or more ion pairs, with multiple ion pairs occurring less frequently (see Table 5). The frequency distribution of ion pairs should therefore be taken into account, or alternatively, the energy distribution (such as that shown in Fig. 27) should be used; however, the mathematics involved are not simple (Harder, 1964). The curves described by equation (5.13) as a function of $\overline{y}$ (and therefore of LET), approach unity asymptotically. On a log-log plot, the initial slope increases as n increases. This trend remains, of course, unaltered if the energy distribution is taken into account. This is shown convincingly in Fig. 30, where the action probability ψ is plotted against LET ($\overline{y}$), for different values of n. The plot is based on the distribution of the number of ion pairs per ion cluster in gases, although values derived earlier than those shown in Table 5 were used (Harder, 1964). The curve for $n = 1$ has an initial slope of 1, which corresponds to the case where ψ approaches unity purely exponentially (see Zimmer, 1961). If the required hit energy, and therefore n, increases then the slope increases too. The result of the combination of equations (5.13) and (5.8) gives the sensitivity curves shown in Fig. 31 a. A comparison with Fig. 31 b, where the energy distribution of Rauth and Simpson has been taken into consideration, shows that the simple expression (5.13) exhibits the essential characteristic: namely, curves with and without maxima.

Fig. 32 demonstrates that the basic idea of the track segment method does apply in practice. The example chosen is the inactivation of haploid

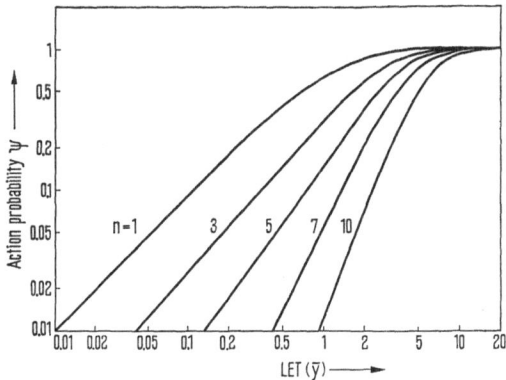

Fig. 30. Action probability ψ as a function of linear energy transfer (LET), for an energy distribution resembling the frequency distribution of ion pairs per ion cluster in gases; n is the least number of ion pairs required for the induction of the test-effect. (Harder, 1964)

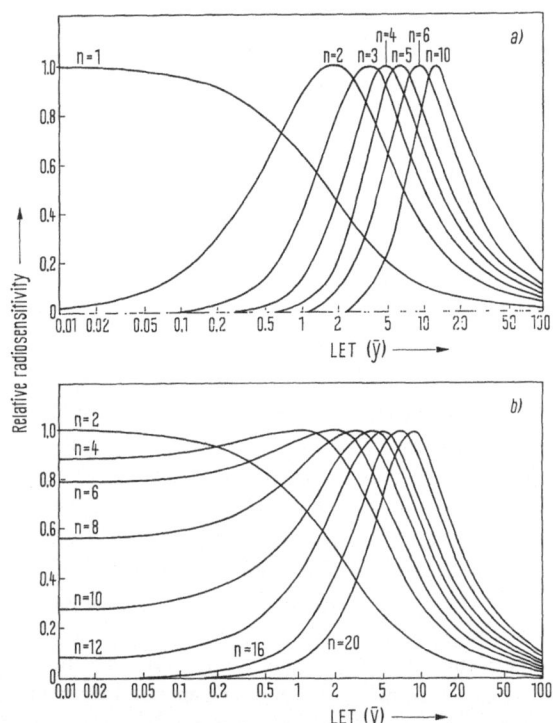

Fig. 31. Relative radiation sensitivity ($1/D_{37}$, normalized to unity) as a function of linear energy transfer (LET). a Calculated from equations (5.8) and (5.13). b Calculated taking into account the frequency distribution of energy losses as measured by Rauth and Simpson (see Fig. 27). (Harder, 1964)

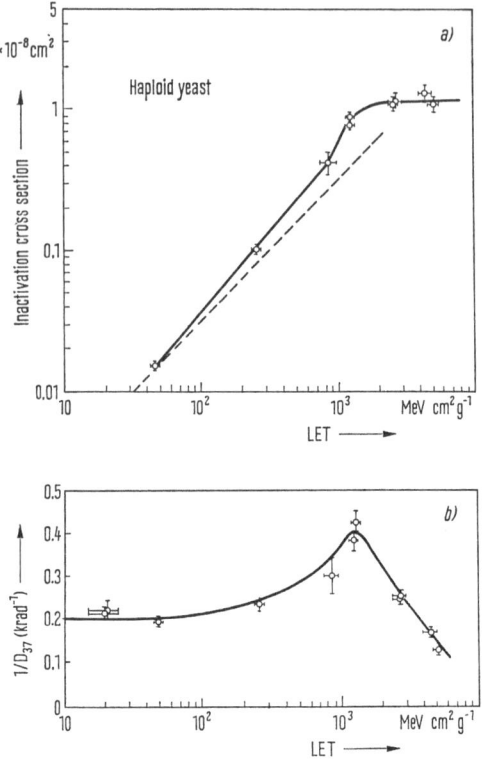

Fig. 32. a Cross section for the inactivation of haploid yeast as function of linear energy transfer (LET) of corpuscular radiation. – – – expected curve for linear LET-dependence. b Radiation sensitivity ($1/D_{37}$) as function of LET for the same experiment. (Sayeg *et al.*, 1959)

yeast cells by charged particles with different LETs. The inactivation cross section is found to increase with LET, initially with a power greater than 1, then to approach a constant value corresponding to the geometrical cross section of the cell nucleus (a). The analogous sensitivity curve (b) reaches a maximum, followed by a hyperbolic decrease in sensitivity at higher LET ("overkill"). The maximum implies that at the corresponding LET value (in this case approximately $1.3 \cdot 10^3$ MeV cm^2 g^{-1}), on average the exact hit energy is transferred per particle passage.

The good agreement between theory and experiment encouraged attempts to calculate the number of necessary energy loss events and the size of the track segment from plots of radiation sensitivity against LET. Fig. 33 shows the results obtained by fitting theoretical curves to experimental results. The inactivation of dried bacterial spores requires an estimated number of at least 8 ions per track segment of 30 Å and per particle. In

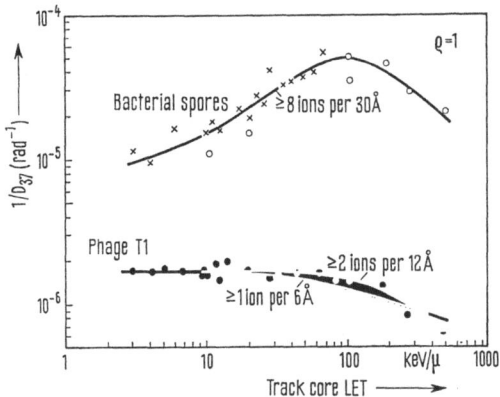

Fig. 33. Radiation sensitivity $(1/D_{37})$ of spores of Bacillus megaterium and of bacteriophage T1 as a function of linear energy transfer (LET), and fitting of the curve expected from the track segment method to the experimental points. (Howard-Flanders, 1958)

contrast, the best agreement for the inactivation of phage T1 is obtained with 2 ion pairs within a track segment of 12 Å. The length of the track segment generally depends to varying degrees on modifying factors, including repair processes. This applies in particular to complex systems, in which the track segment is therefore unlikely to have a real physical meaning. In contrast, the track segment method would be expected to apply in the biologically relatively simple T1-phages. As it is now known that DNA is the primary radiation sensitive structure of the phages, the track segment of 12 Å should be comparable with the average diameter of 17 Å of the DNA double helix. Although the agreement between these two values is in itself remarkable, it is found in addition that at least two ion pairs are necessary per particle passage across the diameter of the DNA molecule, to produce inactivation. It therefore seems reasonable to assume that each DNA strand must receive one ionization per particle passage. This is the first indication that the inactivation event is a double strand lesion in DNA. In fact, double strand breaks are found to be particularly important lethal events in phages of the T-series; this point will be considered further in Chapter 12.3.

The strength of the track segment method, which is essentially a consequence of the fact that a hit is considered to occur after a minimum rather than a mean energy loss event, is further emphasized by a comparison with the target molecular weight determinations for the γ-inactivation of phage T1, using equation (5.5). According to Table 16 (see Chapter 12.3) this amounts to about $1/30$th of the true molecular weight, which shows that although the track segment method does not give the size of the target

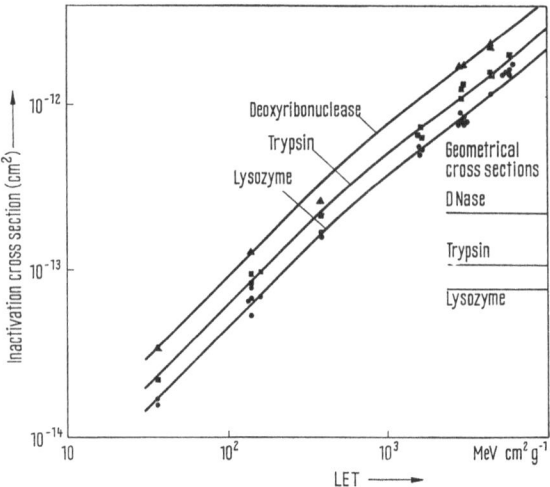

Fig. 34. Inactivation cross sections of deoxyribonuclease, trypsin and lysozyme as a function of linear energy transfer (LET), and a comparison with the geometrical cross sections of the enzyme molecules. (Brustad, 1961)

directly, it can be used to estimate the diameter of the sensitive structure, but only with certain reservations that will now be discussed.

A major inaccuracy that has been committed, results from the imprecise use of the term LET, since its definition based on the Bethe-Bloch equation (4.10) is idealized. So far, the track has been considered as infinitely thin, and it has been assumed that the primary ionizations occur only along this particle track. In reality, however, with increasing LET more delta-rays are produced and these are emitted preferentially in directions perpendicular to the track of the particle, thus resulting in "fatter" tracks. Although the definition of LET includes other idealizations, the neglect of transverse energy transfer is by far the most serious and leads, particularly with small objects, to the breakdown of the target theory in the form used so far. It is observed, for example, that the inactivation cross section of small objects, such as enzyme molecules and phages, does not converge to a constant value with increasing LET, but apparently reaches very high values (see Fig. 34). An attempt has been made to eliminate the influence of delta-rays by the introduction (Harder, 1964) of the "track core" LET (see abscissa in Fig. 33). This procedure, also known as δ-ray correction, is not adequate from the standpoint of exact research because it is somewhat arbitrary, since the extent of the correction generally depends on the actual object size. This procedure will, therefore, not be discussed further (a detailed description is given by Lea, 1946); instead a theoretical approach will be considered that takes the transversal extension of the particle tracks into account.

64

b) The Theory of Butts and Katz

This theory (Butts and Katz, 1967) is based on the following ideas and assumptions:

1. The reactions of biological systems to γ-radiation represent their response to a statistical dose-distribution of δ-rays.

2. The δ-doses (D_δ) which are delivered perpendicular to the track of a heavy ion, can be treated as being statistically distributed over the surfaces of coaxial cylinders, at distance x from the track and of infinitesimal thickness dx.

3. The range of delta-rays may be large relative to the average diameter of small objects, which therefore have to be considered as point objects. The object size is implicit in the D_{37} for γ-radiation (D_{37}^γ), as is implied in point (1).

4. According to equation (5.6) the action cross section σ is equal to the probability that the passage of a particle through a unit area will produce the test effect. Therefore the cross section σ is also (according to the definition of the probability) equal to the quotient of the number of objects hit per particle passage, and the total number of objects present per unit area.

The following expression satisfies these conditions:

$$\sigma = 2\pi \int_0^\infty x\, dx\, (1 - e^{-D_\delta(x)/D_{37}^\gamma}) \qquad (5.14)$$

i. e. all objects on a circular ring of infinitesimal thickness, are multiplied by their corresponding inactivation probabilities (the expression in brackets), and then integrated over all rings. The real problem is thus reduced to a calculation of $D_\delta(x)$, i. e. the dose of delta-rays at a distance x from the track of the primary particle. A short sketch of the principle of this method can be given, without going into detail. The starting point is the energy distribution of delta-rays as a function of the characteristic data of the primary particle initiating them. As the distance x enters equation (5.14), the energy distribution has to be converted into a range distribution. Since there is an upper energy limit for δ-rays, given by the principles of collision interactions, the integral in equation (5.14) has only to be extended to a finite range a. The expression finally obtained is:

$$\sigma = 2\pi \int_0^a x\, dx \left\{ 1 - \exp\left[-\frac{b\, z^{*2}}{2\pi x\, \beta^2\, D_{37}^\gamma} \left(\frac{1}{x} - \frac{1}{a} \right) \right] \right\}, \qquad (5.15)$$

where:

$b \quad = \quad 1.36 \cdot 10^{-7}\, \text{ergs/cm}$

$a \quad = \quad 1.246 \cdot \dfrac{m^2 c^2 \beta^2}{1 - \beta^2}$

$\beta \quad = \quad v/c$

$v \quad = \quad$ the speed of the particle

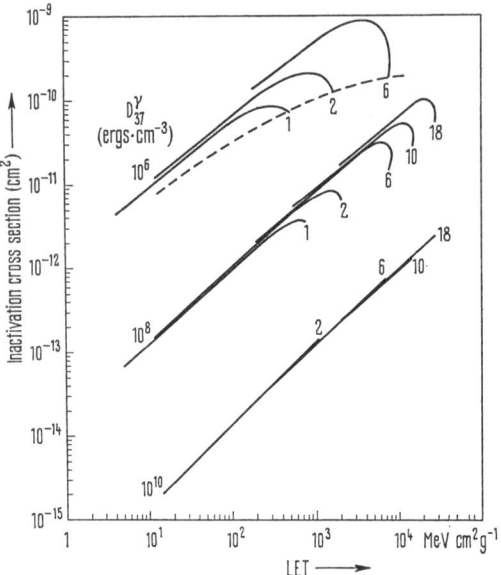

Fig. 35. Action cross section as a function of linear energy transfer (LET), calculated from equation (5.15) for different values of D_{37}^{γ}. (1 erg cm^{-3} = 10^{-2} rad for $\varrho = 1$ g cm^{-3}). The LET scale refers to protons. The numbers on the curves specify the charge of the incident particle (Butts and Katz, 1967)

c = the speed of light

m = the mass of an electron

z^{*} = the effective particle charge as given by equation (4.12).

A detailed discussion of equation (5.15) is outside the scope of this book; however, it is remarkable that the LET does not appear in this formula, instead charge and velocity enter it separately. This automatically takes into account the facts that LET is a function of both these parameters and that it is quite possible that two kinds of particle having the same LET may, dependent on their relative charges and velocities, give different values for σ. This is convincingly demonstrated in Fig. 35, where the family of curves obtained from equation (5.15) is plotted for different values of the ionic charge z and of D_{37}^{γ}. The LET in this plot has been calculated from stopping-power tables for protons. First of all, it can be seen that in the low LET region, σ is proportional to the LET and independent of z. The LET range over which this statement is valid increases for less sensitive objects, i. e. for larger D_{37}^{γ}. In contrast, at high LET and especially with sensitive objects there is a pronounced dependence of the action cross section on the particle velocity, which leads to a sharp reduction in the action cross section. In order to explain these effects, it is necessary to recall that for a given particle the LET passes through the Bragg-maximum at a certain speed (energy), which for protons is slightly below 10^3 MeV cm^2 g^{-1} ($=100$ keV/μ) (see Fig. 20). The σ-curves for protons ($z=1$) therefore end below 10^3 MeV cm^2 g^{-1} (as shown in Fig. 35). The decrease in the action cross section in this region is caused by the reduction at low energies of the range of δ-rays (a in equation 5.15), and thus of the effective diameter of the particle track to dimensions comparable with those of the object, so that the number of objects inactivated per particle passage, and therefore σ, decreases. Similar considerations apply to particles carrying a higher charge. If in the determination of the σ-LET curves for objects of various sizes, the LET could be chosen in such a way that the maximum range of the delta-rays does not exceed the particle diameter, then σ would be expected to converge to the geometrical cross section, and the track segment method would retain its validity. The probability that these conditions will be fulfilled is obviously greatest below the Bragg-maximum (i. e. at the endpoints of the branches of the family of curves associated with a given D_{37}^x value), and especially with the "large" objects of the highest family of curves. If the endpoints of the branches are connected, a clearly convergent σ-curve is obtained (Fig. 35, interrupted curve). Moving to smaller objects, i. e. higher D_{37}^x values, the effective track cross section is nearly always larger than the object cross section; the curve obtained by connecting the endpoints of these branches does not converge.

In practice, ions of different charge and speed are used to obtain the various LET values. The bends in the branches of the curves in Fig. 35 are therefore usually masked, and are at best reflected by some more or less pronounced irregularities in the shape of the measured σ-curves. This is shown in Fig. 36, where the inactivation cross section for phage T1 obtained by experiments and from equation (5.15), making use of the parameters of the ions and the measured D_{37}^x of $5.7 \cdot 10^5$ rad, is plotted. Apart from the good agreement between theory and experiment, it can also be seen that equation (5.15) describes correctly the discontinuity in the inactivation cross section caused by the transition from fast oxygen ions ($\beta = 0.133$) to slow carbon ions ($\beta = 0.073$). Although the LET increases in this interval, the inactivation cross section remains practically constant. Strictly speaking, the plotting of inactivation cross section against LET thereby loses its meaning, as at a given LET the value of σ can be altered by the varying of the charge and velocity. The plotting of radiation sensitivity against LET, according to the theory of Butts and Katz, leads to curves without maxima which also show the same ambiguity as the σ-curves.

It has thus been demonstrated that in the transition to small molecular "targets", the track segment method, which strictly speaking applies only to infinitely thin particle tracks, i. e. primarily to large objects, has to be replaced by a theory that takes the radial distribution of δ-ray doses into

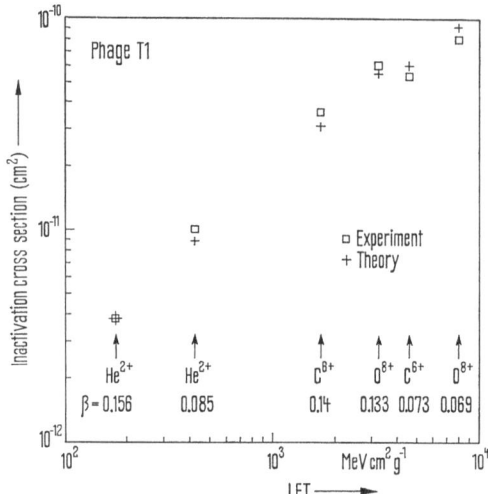

Fig. 36. Cross sections for the inactivation of bacteriophage T1 as function of the linear energy transfer (LET) of the ions used. □ Experimental values (Fluke *et al.*, 1960.) + Calculated values using equation (5.15) with $D_{37}^{\gamma} = 570$ krad and $\beta = v/c$

account. Fig. 35 has been used to demonstrate that there is a continuous transition between the two descriptions, and that for small objects the term LET is ambiguous and, therefore, not a meaningful measure of energy transfer and radiation quality.

5.4. Relative Biological Effectiveness

In conclusion, an application of the target theory will now be considered: the introduction of the concept of relative biological effectiveness for different kinds of radiation. A common problem in radiation biology is to assess the effectiveness of one kind of radiation relative to some "reference" radiation. The difference in effectiveness is taken into account by a factor R which stands for the "relative biological effectiveness" (RBE). R is defined as the ratio of the doses of the radiations being compared which produce the same biological response under identical conditions:

$$R = \left[\frac{D\gamma}{D}\right]_{\text{identical effect}} . \tag{5.16}$$

Gamma-rays are commonly used as the reference radiation, with their dose appearing in the numerator. This definition is necessarily idealized (and does not, for example, apply in the event of intensity dependence) but it is usable in so far as it is possible to determine the RBE from dose-response curves. Of course, a necessary condition is that the quantity R is independent of dose, i. e. that the types of radiation to be compared have the same action kinetics. This restricts the meaningful application of the definition fairly strongly. The treatment of exponential dose-response curves is particularly simple, leading to:

$$R = D^y_{37}/D_{37} \, . \tag{5.17}$$

If the test radiation consist of charged particles, then the D_{37} can be expressed according to equation (5.8). This leads to

$$R = \frac{\varrho \cdot D^y_{37}}{L} \, \sigma \, (L) \, . \tag{5.18}$$

This means that the RBE depends on the LET of the test radiation qualitatively in the same manner as the radiation sensitivity. In other words, depending on the LET-dependence of σ, R is either initially a constant equal to unity (as for example, in the theory of Butts and Katz) and decreases at high LET values, or alternatively it goes through a maximum with a value greater than 1, and then decreases. Of course, R will usually vary for different test effects. For the requirements of radiation protection, the specific quantity R is replaced by the "quality factor", in which the estimate of risk is based upon various responses of the body. For this reason quality factors, in contrast to R, are not directly measurable but represent, in a manner of speaking, empirical values.

References

Brustad, T.: Radiat. Res. 15, 139 (1961).
Butts, J. J., Katz, R.: Radiat. Res. 30, 855 (1967).
Fluke, D. J., Brustad, T., Birge, A. C.: Radiat. Res. 13, 788 (1960).
Harder, D.: Biophysik 1, 225 (1964).
Howard-Flanders, P.: In: Advances in biological and medical physics, Vol. VI. Eds.: C. A. Tobias and J. H. Lawrence. New York: Academic Press 1958, p. 553.
Lea, D. E.: Actions of radiations on living cells. Cambridge: University Press 1946.
Pollard, E. C.: Rev. Mod. Phys. 31, 273 (1959).
— Whitmore, G. F.: Science 122, 335 (1955).
Rauth, A. M., Simpson, J. A.: Radiat. Res. 22, 643 (1964).
Sayeg, J. A., Birge, A. C., Beam, C. A., Tobias, C. A.: Radiat. Res. 10, 449 (1959).
Zimmer, K. G.: Studies on quantitative radiation biology. Edinburg-London: Oliver & Boyd 1961.

Chapter 6. Direct and Indirect Action of Radiation

In the previous chapters, an attempt has been made to draw conclusions as to the nature of the radiation lesions by applying formal physical and mathematical procedures to the interpretation of dose-response curves, and by investigating the LET-dependence of radiation sensitivity. It was assumed that such a thing as a well defined target really exists. Basically, this assumption seems to be justified, since the formal definition of target used in the hit theory was so general that difficulties are only encountered when attempts are made to identify the target with sensitive biological structures. However, as this is the main theme of the target theory, it is necessary to consider the extent to which realistic targets can be obtained from dose-response curves. The concept of a target does not make any allowance for damage from the "outside", which is the rule rather than the exception. If the concept is to be retained, then the fraction of the energy contributed from outside to the target must be determined. This leads to the concepts of direct and indirect effect as outlined in Chapter 1.3. Such a classification is meaningful only at the molecular level, where the chances of distinguishing between these two effects are greatest. If the absorption of radiation occurs in the molecule in which the lesion appears, then this is the *direct action of radiation*, while with *indirect action* the absorption of the radiation energy and the response to this energy occur in different molecules. This definition is considerably more rigorous than the one used in the past, in which the irradiation of dry systems was considered as direct, while the indirect effect was considered to occur predominantly in the presence of water. It is now known that even in the irradiation of the purest dry substances under high vacuum, indirect action contributes to the observed damage. The indirect effect is generally attributed to the attack of free radicals, produced by the radiation in the vicinity of the molecule under consideration. However, intermolecular energy transfer is also part of this phenomenon (see Fig. 2), while intramolecular energy transfer is without doubt a mode of the direct action.

6.1. The Direct Effect

It was shown in Chapter 4 that the absorption of ionizing radiation in matter occurs via a number of different primary processes such as excitation, ionization, elastic nuclear collisions, etc. As direct action is a result

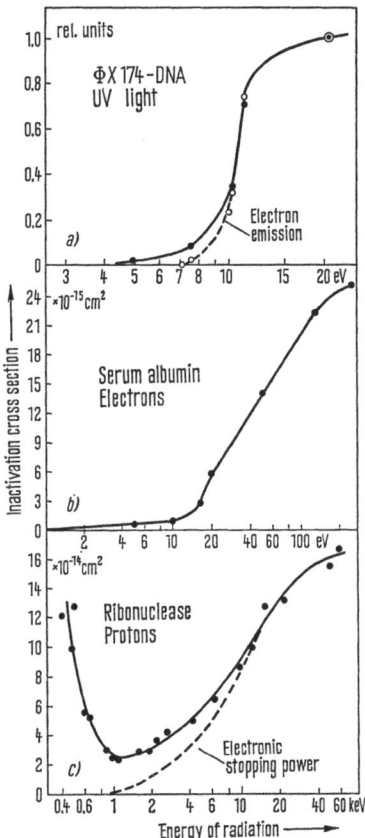

Fig. 37. Biological effectiveness of the various primary processes of energy absorption. a Comparison of the inactivation cross section of infectious ΦX 174-DNA with the electron emission from thick layers of calf thymus DNA, using vacuum-ultraviolet radiations of various quantum energies (Berger, 1969). b Cross section for the inactivation of bovine serum albumin by electrons of different energies (Hutchinson, 1960). c Cross section for the inactivation of ribonuclease by protons of different energies (Jung, 1965)

of these primary processes, the biological effectiveness of the various absorption processes must now be considered. Normal ionizing radiation is unsuitable for the independent investigation of *excitations* and *ionizations*, as both of these primary processes always occur simultaneously (Chapter 4.4). However, short wavelength ultraviolet radiation, known as vacuum UV, can be used for the investigation of excitations and ionizations, as well as for the region of transition from excitations to ionizations. An example of the effectiveness of vacuum UV as a function of wavelength is given in Fig. 37 a, which shows the inactivation cross section of the infectious DNA of phage ΦX 174 (see Chapter 11.2) as a function of the energy of the incident UV quanta. In addition, the probability per quantum of electron emission from the DNA is plotted. The values obtained in these two experiments were set equal to unity at 58.4 nm (corresponding to 21.2 eV). Above 10 eV both curves coincide, indicating that the inactivation is mainly due to ionization. Both effects decrease with decreasing quantum energy and below 7.1 eV no electron emission is observed. The inactivation in this energy region is therefore caused by excitations.

Similar results are obtained for the inhibition, by electrons of various energies, of the immunological action of bovine serum albumin, (Fig. 37 b). A significant effect occurs only above 10 eV, and it increases at higher energies in the same manner as the ionization probability. From these experiments it may be concluded that, in general, amounts of energy below 10 eV have little radiation biological significance, i. e. only rarely will the excitation of the lower energy states have any biological consequences.

The curve of Fig. 37 c shows the cross section for the inactivation of ribonuclease by protons of various energies. Above 10 keV, the inactivation curve shows the same dependence on energy as the ionization curve of the protons (see Fig. 20). The curve does not approach zero at low proton energies as does the ionization probability (Fig. 37 c, broken curve), but increases sharply. This increase is due to *elastic nuclear collisions,* the frequency of which increases with decreasing energy (see Chapter 4.3). The cross section for the inactivation of ribonuclease by elastic nuclear collisions reaches approximately the same magnitude as that for the inactivation by protons with maximal ionization density (Bragg maximum: 60—100 keV; see Fig. 20). The inactivation of DNA by elastic nuclear collisions is similar to that of ribonuclease (RNase) (Jung and Kürzinger, 1969). It is a characteristic of elastic collisions that their action is not modified by protective agents or by low temperature (Jung, 1966).

These examples may suffice to illustrate the effectiveness of the various direct radiation effects. They show that the occurrence of an ionization or nuclear collisions within a macromolecule has a high probability of causing inactivation, while excitations have a much lower biological effectiveness.

6.2. Indirect Effect in Solutions

The distinction between indirect effects in solution and in the dry state is made predominantly for historical reasons, but also to give a better grouping of the multitude of results and phenomena. The distinction is historical because in the past it was assumed that indirect action is possible only in the presence of water, but this assumption has since been proved to be incorrect. It is now known that the indirect effect in solutions, and also in the dry state, is mainly due to the action of small diffusible radicals released by the action of radiation. A study of the indirect effect in solutions is, therefore, essentially an investigation of the production of these radicals by radiation, and of their reactions with biologically important macromolecules.

a) Radiolysis of Water and the Primary Reactions of the Products

Exposure to ionizing radiation leads to the formation of highly reactive species in the most diverse solvents. However, only the example of water will be considered in any detail, since it occupies a position of special importance because of its presence in practically all biological systems. The reactions of other substances can be found in textbooks of Radiation Chemistry (e. g. Spinks and Woods, 1964).

As the radiolysis of water is by no means a trivial problem, a short and simplified summary will have to be given. The absorption of radiation energy leads primarily to the ionization of water molecules (ionization potential 12.56 eV):

$$H_2O \rightarrow H_2O^+ + e^- . \tag{6.1}$$

The positive ion formed in this process can lead to the formation of OH radicals by:

$$H_2O^+ \rightarrow H^+ + OH^. \tag{6.2}$$

and, the reaction of electrons with water molecules to the production of H radicals (hydrogen atoms):

$$e^- + H_2O \rightarrow OH^- + H^. . \tag{6.3}$$

The H$^.$ and OH$^.$ radicals are formed not only by these greatly simplified reactions, but also directly by excitation (about 7 eV) and dissociation of a water molecule:

$$H_2O \rightarrow H_2O^* \rightarrow H^. + OH^. . \tag{6.4}$$

The electrons released in the primary process (equation 6.1) are of special biological interest, as they may polarize water molecules in their vicinity, and thereby become stabilized to what are known as hydrated electrons (e_{aq}^-). In this "long-lived" mode, they can diffuse over considerable distances, and react effectively with dissolved biological molecules. The biologically important radiolytic products, H$^.$, OH$^.$, and e_{aq}^- are referred to as *water radicals*.

Although a pH-dependence of the *yields of water radicals* might be expected on the basis of these reactions, it is only significant at very high or very low pH. In the region between pH 3 and pH 10, the G-value (i. e. the number of species produced per 100 eV of absorbed radiation energy) for sparsely ionizing radiation is approximately (Buxton, 1966):

$$G_{e_{aq}^-} = 2.3; \quad G_{H^.} = 0.6; \quad G_{OH^.} = 2.3 .$$

The radiolysis of water also leads, via the recombination of H$^.$ and OH$^.$, to the formation of the molecular secondary products H_2, H_2O and

H_2O_2. Although these reactions in general cause no significant inactivation of dissolved materials, they nevertheless are important in the case of high LET radiation, where the local concentration of water radicals is so high that their recombination, and therefore the formation of H_2, H_2O and H_2O_2 is favoured at the expense of radical products. This increase in the yield of H_2 and H_2O_2 with LET, and the consequent decrease in indirect action, can be observed experimentally. For example, Zimmer and Bouman (1944) found that dissolved tyrosine is decomposed more efficiently by X-rays than by alpha-particles.

The radicals $H^{\cdot}$ and $OH^{\cdot}$ can act as reducing or oxidizing agents on the biological molecules MH. Typical primary reactions are radical addition or the abstraction of hydrogen:

$$MH + H^{\cdot} \rightarrow MH_2^{\cdot} \qquad (6.5)$$

$$MH + OH^{\cdot} \rightarrow MHOH^{\cdot} \qquad (6.6)$$

$$MH + H^{\cdot} \rightarrow M^{\cdot} + H_2 \qquad (6.7)$$

$$MH + OH^{\cdot} \rightarrow M^{\cdot} + H_2O. \qquad (6.8)$$

Similar reactions may be formulated with e_{aq}^{-} . The organic radicals $M^{\cdot}$, $MH_2^{\cdot}$ and $MHOH^{\cdot}$ produced by the reactions (6.5) to (6.8) can undergo intramolecular changes, and via various modes of reaction reach a state of irreversible damage.

b) Kinetics of the Indirect Action of Radiation

The kinetics of the reactions between dissolved substances and the water radicals produced by radiation will now be considered. It will be assumed that the number of radicals produced is proportional to the dose, and that they do not interact with each other. Two situations can then be distinguished:

1. Water radicals can only react once with a dissolved molecule, which is changed to such an extent that a second reaction is not possible. Under these conditions, it is to be expected that the number of damaged molecules will increase linearly with dose. The "survival rate" is therefore given by (see Fig. 38):

$$N/N_0 = 1 - kD. \qquad (6.9)$$

Such kinetics are frequently observed for small molecules, e. g. the oxidation of dissolved $FeSO_4$ by the OH radical. This reaction

$$Fe^{2+} + OH^{\cdot} \rightarrow Fe^{3+} + OH^{-} \qquad (6.10)$$

is of special importance in dosimetry (Fricke dosimetry). Extrapolation of the linearly plotted dose-response curve to $N/N_0 = 0$ gives the dose D_0, this being the dose at which all dissolved molecules have been converted. A consequence of the postulate that a molecule can react only once with

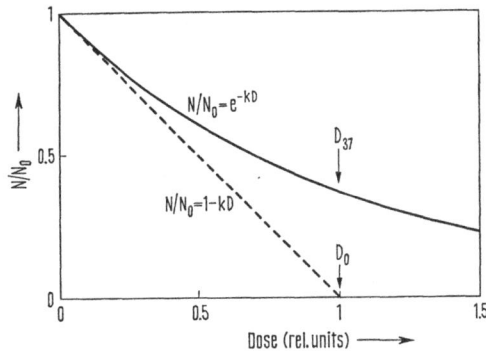

Fig. 38. Inactivation kinetics for irradiation in dilute aqueous solutions

a radical is that, at the dose D_0, the number of water radicals having reacted equals the total number of dissolved molecules, and this can be used to determine the energy necessary for the conversion.

2. In the second case, it is assumed that the radiation-induced radicals react at random with the dissolved molecules, with equal probabilities of reacting with changed or unchanged molecules. It is obvious that this applies with most biological molecules, as these relatively large molecules are not changed by a water radical to such an extent that there is no possibility of a second or third attack by a radical. The considerations of the hit theory, described in Chapter 2, apply here. If the test effect is caused by a single effective event, then an exponential dose-response curve is obtained, described by (see Fig. 38):

$$N/N_0 = e^{-kD} . \tag{6.11}$$

The factor k (which is used in the conventional sense, standing for the reciprocal of D_{37}), does not, of course, represent the size of the biological target, since it depends on concentration, as will now be shown. As at the $37^0/0$-dose the number of attacks equals the number of molecules present (see Chapter 2), the D_{37} can be used to calculate the amount of energy per inactivating event. Exponential inactivation curves are obtained as a rule in the irradiation of enzymes and single-stranded DNA in solution.

Shoulder curves are, however, sometimes observed with irradiated solutions, e. g. for the inactivation of phage T7 in suspension (Fig. 92). Whether or not such curves are obtained depends very much on the nature of the radiation damage being observed, and they are an indication of the possibility that certain types of damage can be caused by "multi-hit processes", or in other words by the accumulation of "sublethal" damage.

c) The Yield of Damaged Biomolecules

Fig. 39 gives some idea of the magnitude of the indirect effects in solution; it shows the dose-response curves for the inactivation of ribonuclease by

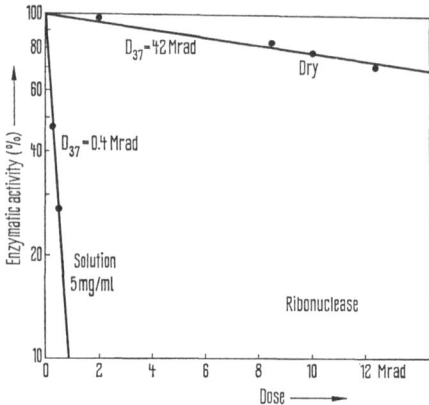

Fig. 39. Inactivation of ribonuclease in the dry state and in aqueous solution by ^{60}Co γ-radiation. (Günther and Jung, 1967; Jung and Schüssler, 1966)

gamma-radiation, in the dry state and in solution (5 mg/ml). The D_{37} values of 42 and 0.4 Mrad, respectively, show that the RNase molecule in dilute solution is approximately 100 times more sensitive than in the dry state. This indicates that, at this concentration, about 99% of the RNase molecules are inactivated by the attack of water radicals, while only 1% are inactivated by the absorption of radiation energy in the macromolecules themselves. Therefore, *in dilute aqueous solutions the indirect action of the radiation is predominant.*

The increased sensitivity in solution could be formally explained in terms of a target, being 100 times larger than for dry RNase, and representing the region around the enzyme molecule from which the absorbed radiation energy may be transferred to the molecule via water radicals. The high sensitivity of RNase in solution is certainly not due to the individual molecules having lower inactivation energies than in the dry state. This is shown by the calculation of G-values, i. e. the number of molecules inactivated per 100 eV of absorbed radiation energy. Using this definition, the G-value can also be presented by:

$$G = Z_m/Z_t, \qquad (6.12)$$

where Z_m is the number of irradiated molecules per gram and Z_t is the number of 100 eV "units" deposited per gram, at the D_{37} (or D_0, depending on inactivation kinetics). In dry systems Z_m is obtained simply by the division of Avagadro's number by the molecular weight MW of the biomolecules:

$$Z_m = 6.022 \cdot 10^{23}/MW, \qquad (6.13)$$

while Z_t is obtained using equation (5.1):

$$Z_t = 6.24 \cdot 10^{11} \cdot D_{37}, \qquad (6.14)$$

76

where D_{37} is measured in rad. The G-value for irradiated dry molecules is therefore given by:

$$G_{\text{dry}} = Z_m/Z_t = \frac{9.65 \cdot 10^{11}}{D_{37} \cdot MW}. \qquad (6.15)$$

For dilute solutions, Z_m is obtained only after the multiplication of equation (6.13) by the fractional weight of the dissolved substance, i. e. by the concentration C. The G-value for irradiated solutions is therefore calculated using the following formula:

$$G_{\text{solution}} = \frac{9.65 \cdot 10^{11} \cdot C}{D_{37} \cdot MW}. \qquad (6.16)$$

Entering the data for RNase from Fig. 39 (dry: $D_{37} = 42$ Mrad; wet: $D_{37} = 0.4$ Mrad, $C = 0.005$; $MW = 13680$) into equations (6.15) and (6.16) gives

$$G_{\text{dry}} = 1.68; \quad G_{\text{solution}} = 0.89. \qquad (6.17)$$

In spite of the large difference in D_{37}, the two G-values are of similar magnitude with the inactivation in the dry state being even more effective than in solution.

d) Concentration Dependence

It might be concluded from the above considerations that the G-values for damaged molecules are essentially independent of concentration. On the other hand, the D_{37} for the inactivation of chymotrypsin increases linearly with concentration in regions of intermediate concentration, as is shown in Fig. 40, and this is also observed with other biomolecules. The fact that the dilution of a solution of macromolecules, (e. g. by 1 : 10) increases the sensitivity by the same factor (i. e. by a factor of 10), is referred to as the *dilution effect*. It is an important criterion for the indirect action of radiation. If, instead of D_{37}, the quantity C/D_{37} (which is proportional to the G-values; cf. equation 6.16) is plotted against the concentration C, a horizontal straight line is obtained at intermediate concentrations (Fig. 40 b); i. e. the energy required for the inactivation of a molecule does not depend on the concentration. This observation, and the occurrence of a dilution effect, can only be understood if it is assumed that a specific number of water radicals are produced per unit dose, which in turn inactivate a certain number of biomolecules. It is apparent that in very dilute solutions the water radicals have to migrate over relatively larger distances before reaching a macromolecule. In extremely dilute systems, some of the radicals will interact or react with impurities, and as a consequence will not be available for the initiation of the test effect, thereby leading to a decrease in the yield (Fig. 40 b). On the other hand, the estimates (6.17) show that the inactivation in the dry state is slightly more effective than in solution.

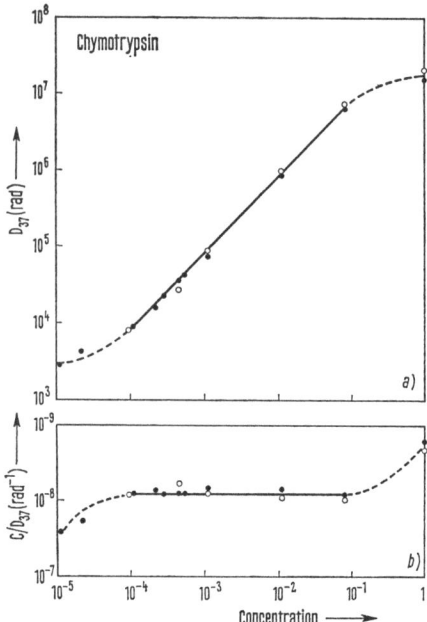

Fig. 40. a 37%-dose (D_{37}) for the inactivation of α-chymotrypsin in aqueous solution with 15 MeV electrons as a function of concentration C. b C/D_{37} as a function of concentration. ● Esterase activity; ○ Protease activity. (Butler *et al.*, 1960)

It may, therefore, be concluded that as the concentration increases, more and more molecules are changed by the direct absorption of radiation, so that the yield at high concentrations increases (Fig. 40 b).

It is therefore obvious that the *absolute number* of molecules inactivated in solutions is a function of dose:

$$\text{solution: } N^+ = f(D). \qquad (6.18)$$

In the range of intermediate concentrations, N^+ is proportional to the dose (see Fig. 40). In contrast, in dry (i. e. predominantly "direct") systems, the *relative* fraction of inactivated units N^+/N_0 is a function of dose as was shown in Chapter 2:

$$\text{dry: } N^+/N_0 = f(D). \qquad (6.19)$$

This implies that if a macromolecule is exposed to a certain dose of radiation D, different concentration dependencies are obtained in the dry state and in aqueous solution, when N^+ or N^+/N_0 are plotted against the concentration (Fig. 41). According to equations (6.18) and (6.19) N^+ in solution and N^+/N_0 in the dry state are independent of the quantity of irradiated material N_0 (Fig. 41, a and b, horizontal straight lines). Alternatively, these

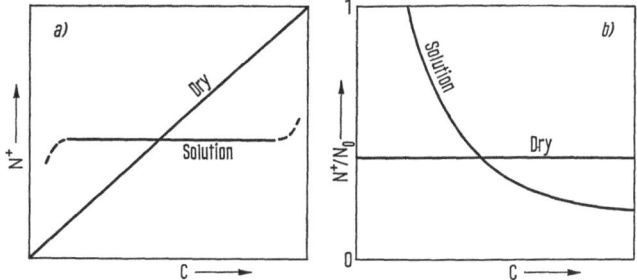

Fig. 41. a Absolute number of molecules changed (N^+) per unit dose as a function of concentration (or quantity) of the irradiated material. b Relative proportion of molecules changed (N^+/N_0) per unit dose as a function of concentration (or quantity) of the irradiated material. (Bacq and Alexander, 1961)

equations can be written as:

$$\text{solution:} \quad N^+/N_0 = f(D)/N_0 \tag{6.20}$$

$$\text{dry:} \quad N^+ = N_0 \cdot f(D). \tag{6.21}$$

Thus, at a given dose, in solution N^+/N_0 decreases with N_0, while in the dry state the number of inactivated molecules, N^+, is proportional to the total number irradiated, N_0 (Fig. 41).

6.3. Indirect Effect in Cells

It is interesting, in this context, to enquire what contribution is made to cellular lesions by the indirect effect. The discussion will be confined to those phenomena caused by the attack of radicals derived from intracellular water. Measurements of enzyme activity, or transformation experiments using DNA from irradiated cells (see Chapter 11.3) are particularly suitable for the study of the intracellular effects. Hutchinson et al. (1957) irradiated yeast cells with different water contents, and recorded the dose-response curves for various enzymes. They found, for example, that invertase is about one-half as sensitive to radiation in dry cells as in the wet cells. Since the sensitivity in dry cells coincides with that of the purified dry enzymes, it was concluded that the contribution of the indirect effect in wet cells is comparable with that of the direct effect. In other cells, the invertase test showed no dependence on water content (Pauly et al., 1966). Transforming DNA extracted from irradiated vegetative cells of Bacillus subtilis, is 4 times as sensitive to the action of gamma-rays or fast electrons as DNA extracted from spores irradiated in the dry state (Tanooka and Hutchinson, 1965; see Fig. 80).

The radiation sensitivity of the cell itself can, of course, be studied as a function of water content. This is shown in Fig. 42 for the inactivation of the bacterium E. coli. D_{37} values between 3.6 and 12.4 krad are obtained

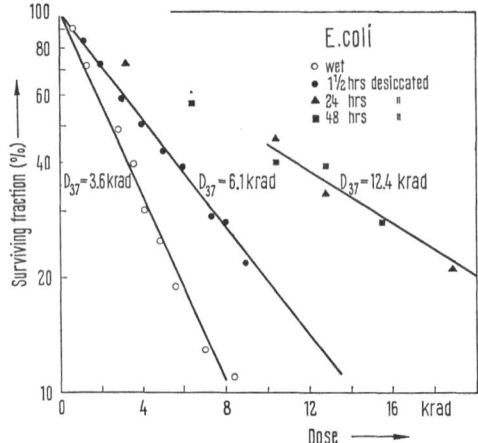

Fig. 42. Inactivation of coli bacteria at different degrees of dehydration. (Bhattacharjee, 1961)

depending on the degree of desiccation (Bhattacharjee, 1961). Similarly, the sensitivity of vegetative cells and dried spores of the same type of bacteria can be compared; in the case of Bacillus subtilis, the colony forming ability of cells is 4 times more sensitive than that of spores (Tanooka and Hutchinson, 1965), and this agrees with the difference between sensitivities of the transforming DNA extracted from these two forms (see above).

Many more examples could be given. They all lead to a similar conclusion, namely, that the intracellular direct and indirect effects are of comparable magnitude, although in a specific case it does, of course, depend on the particular cell function that is tested.

6.4. Indirect Effect in the Dry State

The above discussion of the relationship between direct and indirect effect in cells is greatly simplified and even, strictly speaking, incorrect, since in the preceding sections it has been assumed that the radiation response observed in dry cells is exclusively due to the direct effect. In fact, however, indirect action occurs in the dry state as well as in solution. The term "dry" has a rather nebulous meaning, when applied to microorganisms. But even when it is used in its strictest sense an indirect effect is still observed: for example, during the irradiation of highly purified dried enzymes in vacuo. However, the basic criteria mentioned above, such as the dilution effect, cannot be applied in this situation.

Formation of Atomic Hydrogen: The attack of atomic hydrogen derived from organic molecules (MH) is probably the most important indirect reaction mechanism in dry systems:

$$MH \rightarrow M^{\cdot} + H^{\cdot} . \tag{6.22}$$

Müller and Dertinger (1968), among others, have shown that hydrogen atoms are released in the irradiation of dried DNA and whole phages: the hydrogen atoms produced at 130 °K appear in the ESR-spectrum as two characteristic lines separated by 506 Oe (Fig. 43).

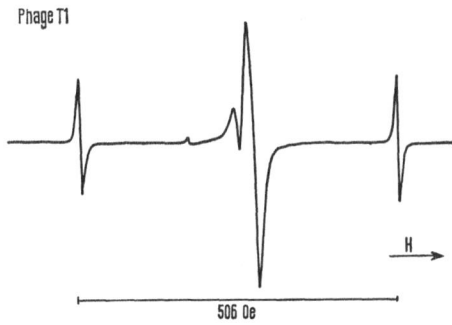

Fig. 43. ESR-spectrum of irradiated T1-bacteriophages at 130° K showing the dublet of atomic hydrogen (splitting: 506 Oe). (Müller and Dertinger, 1968)

Another indication of the formation of atomic hydrogen in dry systems is the fact that the gases formed during irradiation of carbohydrates (e. g. Swallow, 1960), amino acids (Sommermeyer *et al.*, 1967), proteins (ten Bosch and Braams, 1963) and nucleic acid constituents (Heitkamp *et al.*, 1968) consist mainly of molecular hydrogen. In addition, various molecular fragments are observed, whose frequency decreases with increasing molecular weight.

Reactions of Atomic Hydrogen. The hydrogen released during irradiation reacts with undamaged molecules, mainly by addition to a double bond, or by the abstraction of a further hydrogen atom (equation (6.5) and (6.7)). The radicals formed in these processes can be identified using ESR-spectroscopy (e. g. in DNA constituents; see Table 12). Similar ESR-spectra can be obtained in many cases by the exposure of dried powdered biological substances to atomic hydrogen obtained from gas discharges (e. g. Heller and Cole, 1965).

Besides this physico-chemical evidence, it is possible to directly examine the damaging action of hydrogen atoms, formed according to equation (6.22), on a biological system. The experimental arrangement shown in Fig. 44, which has some resemblance to a plate condenser, was used by Jung and Kürzinger (1968) to produce atomic hydrogen. The samples to be inactivated face a polyethyleneterephthalate foil, from which hydrogen atoms are released by a beam of 2 MeV protons, and then diffuse to the samples and react with them. It is found, for example, that DNA of the phage ΦX 174 (Fig. 44), RNase, and even whole T1-phages are inactivated exponentially. The temperature dependence of these reactions will be discussed in the following chapter.

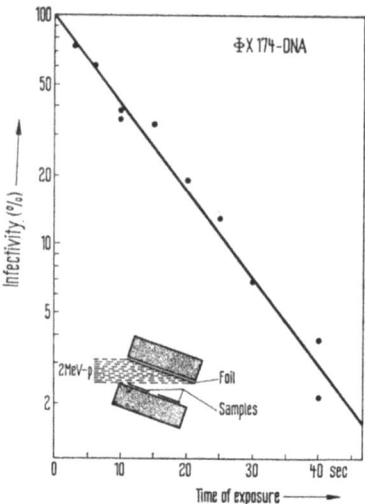

Fig. 44. Inactivation of infectious ΦX 174-DNA by atomic hydrogen produced by irradiation of a foil of polyethyleneterephthalate with 2 MeV protons. (Jung and Kürzinger, 1968)

It is difficult to estimate the magnitude of the indirect effect in dry systems, as quantitative data are not available, either for the production or for the reaction of hydrogen atoms in biological materials. Nevertheless, it is possible to give some idea of its magnitude with the aid of the temperature effect, which represents the most important criterion of indirect action in dry systems (see Chapter 7). If it is assumed that the temperature-dependent part of the inactivation cross section (see Fig. 51) reflects indirect action, while the temperature-independent part is associated with direct action, then the relative contributions of the direct and indirect effects are comparable.

Intermolecular Energy Transfer. Attack by radiation-induced hydrogen atoms is certainly the predominant and most important mechanism by which radiation energy is transferred to an undamaged biomolecule in the dry state. However, other mechanisms of the indirect action of radiation cannot be excluded. The phenomenon of intermolecular energy transfer has to be discussed in this context; this energy transfer occurs, for example, in plastic scintillators. Most of the ionizing radiation is absorbed in the major constituent of the plastic scintillator (usually polystyrol), and some of the energy is then transferred to additives present at relatively low concentrations (e. g. terphenyl), causing them to become luminescent. Intermolecular energy transfer processes occur naturally in many biological systems, for example, in algae, which utilize the light energy absorbed by suitable pigment molecules; in most cases the quantum yield is very much larger

82

than in scintillators. However, it has not yet been shown whether this process plays any part after the irradiation of highly purified, biologically important macromolecules, although intermolecular energy transfer has been investigated in irradiated molecular mixtures of, for example, proteins and small sulphur-containing molecules. Temperature-dependent changes in the ESR-spectra of such mixtures are observed, which indicate a transfer of the radical state to the sulphur-containing molecules (e. g. Gordy and Miyagawa, 1960; Henriksen et al., 1963). However, it is possible that this spin transfer could be caused by diffusible radicals, in which case the results could be explained in terms of the hydrogen donation hypothesis (equation 6.25), which will be discussed in the section on protective agents. A detailed description of the problems of intermolecular energy transfer is given by Phillips (1966).

The special case of *frozen solutions* will be considered in conclusion, as an extension of these considerations of the indirect effect. The radiation sensitivity of dissolved trypsin has a discontinuity at the freezing point (Fig. 56), the reduction of sensitivity being, no doubt, caused by inhibition of the diffusion of radiation-induced water radicals. However, below the freezing point, the radiation sensitivity varies with temperature, and in addition is always greater than for irradiation in the dry state. This shows clearly that even in ice the indirect effect by no means completely disappears, although it is not quite clear what indirect processes are involved. Finally, it should be mentioned that this discussion of the direct and indirect effect will be extended in Chapters 7 and 8, i. e. in the treatment of the temperature and oxygen effects.

6.5. Protective and Sensitizing Agents

The elucidation of the mechanisms of action of protective and sensitizing agents is one of the most important problems of molecular radiation biology. Whilst the possible importance of radiation-protective agents is obvious, it may be necessary to point out that there are important applications of sensitizing agents, e. g. in the selective sensitization of tumours in radiotherapy. The actions of substances which modify the radiation sensitivity are complex and varied; consequently, only a superficial understanding has been obtained in spite of numerous experiments. This section will not, therefore, deal with the confusing multitude of experimental results, but will be restricted to a short summary of some of the basic facts of the action of these substances. It is important that the terms protection and sensitization should be used with great care. For example, the term sensitization may refer to a blockage or inhibition of a naturally occurring protective mechanism, or alternatively to a true sensitization resulting from intermolecular energy transfer or from an increased yield of free radicals, i. e. a general enhancement of the indirect effect. For example, dry trypsin is more sensitive to irradiation in the presence of dextran, ribose or lactose (Tobias *et al.*, 1960). Exposure to visible or UV light may also cause sensitization, via "photodynamic" effects; more information about this is contained in a paper by Spikes and Straight (1967). Because of the difficulty in defining the general principles of the action of sensitizing agents, this topic, although interesting, cannot be considered further. Consequently, the following discussion will concentrate on the most important aspects of the action of protective agents.

At the molecular level, two kinds of protective action may be distinguished: competition and restitution. (The term "repair" often used in connection with chemical protection will be exclusively reserved for the enzymatic repair processes discussed in later chapters.) *Competitive protection* refers to the ability of certain chemical compounds to compete with the biomolecules for the diffusible radicals. The magnitude of the resultant protection depends on the relative concentrations of the two competing molecules, and also, of course, on their reaction constants with the radicals. Protective agents of this kind are known as radical scavengers, and will be considered in more detail later. A characteristic of competitive protection is that the contribution of the indirect effect is reduced. In contrast, in the other important protective mechanism, that of *restitutive protection*, the protector does not alter the number of primary lesions. Instead, some of the damage is restituted by the protective agent, so that an overall protection is observed.

a) Protection in the Dry State

A particularly important group of protective agents that are also effective in dry systems are those chemicals containing a sulphydryl group (general formula RSH) or a disulphide bond (general formula R_1SSR_2). Of these, cysteine, cystine, cysteamine, cystamine, thioglycol and glutathione have been examined in detail. As an example of protective action in a dry system, the dose reduction factor (i. e. the ratio of the 37%-doses in the presence and absence of a protective agent) for the inactivation of ribonuclease is plotted against the concentration of the protective agent cystamine (Fig. 45). The maximum dose reduction factor of approximately 1.8 is obtained with a weight ratio (cystamine to RNase) of about 0.8. Qualitatively similar curves are obtained for other agents in different systems.

The protective action in a dry system may be either competition or restitution. The possibility of competitive protection is reflected in Braam's hypothesis (1963) according to which the hydrogen atoms formed by equation (6.22) react with the protective agent:

$$H^{\cdot} + RSH \rightarrow RS^{\cdot} + H_2 \tag{6.23}$$

$$H^{\cdot} + R_1SSR_2 \rightarrow R_1S^{\cdot} + R_2SH, \tag{6.24}$$

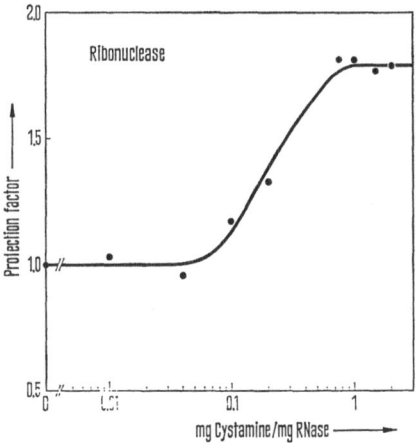

Fig. 45. Protection factor (ratio of D_{37}-values with and without protective agent) for the inactivation of dry ribonuclease by 2 MeV protons as a function of the amount of cystamine added. (Jung, 1966)

thereby preventing the damage of further biomolecules by the mechanisms of equations (6.5) and (6.7). To allow for the possibility of restitution in dry systems, Alexander and Charlesby (1955) and Howard-Flanders (1960) proposed a hydrogen donation mechanism, in which a damaged biomolecule M˙ is restituted by the transfer of a hydrogen atom from a sulphydryl-containing protector:

$$M˙ + RSH \rightarrow MH + RS˙ . \qquad (6.25)$$

Restitution by disulphide compounds would, however, be difficult to explain in the absence of a reducing agent.

A decision in favour of one or the other of these mechanisms is very difficult, as both mechanisms may play a part. One interesting implication of Braam's hypothesis is that the maximum dose reduction factor obtainable is 2, and this is often observed experimentally (e. g. Fig. 45). It arises from the fact that for each radical M˙ produced, one hydrogen atom is released (equation (6.22)) which can in turn react with another molecule MH (see equations (6.5) and (6.7)). If the hydrogen radicals are scavenged completely (equations (6.23) and (6.24)), then the yield M˙ is halved, giving a dose reduction factor of 2. In contrast, equation (6.25) allows for the possibility of unlimited protection. Nevertheless, a finite dose reduction factor can be introduced into the hypothesis of Alexander, Charlesby and Howard-Flanders, when the fact that equation (6.5) also applies in the dry state is taken into consideration. It is then sufficient to assume that only M˙, and not $MH_2˙$, is reverted by SH compounds to its original functional state.

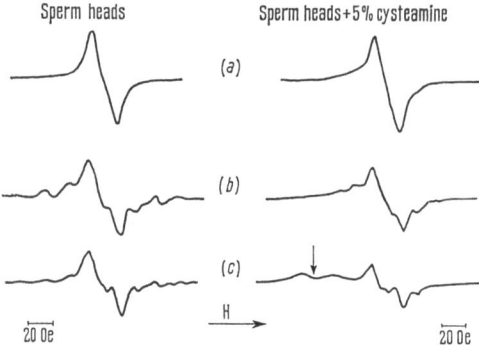

Fig. 46. ESR-spectra of herring sperm DNA with and without the addition of cysteamine, after irradiation at 77 °K. a Measurement at 77 °K. b Measurement after warming to room temperature. c Measurement after storage for 6 days at room temperature. (Ormerod and Alexander, 1962)

Equations (6.23), (6.24), and (6.25) all lead to the formation of sulphur radicals RS˙. Radicals of this type can be detected using ESR, and are recognized by the typical tailing-off of the ESR-spectrum at low field-strengths. The formation of these radicals will be followed in an example which seems to support reaction (6.25). If herring sperm, which are 65% DNA, are irradiated in vacuo at 77 °K, then identical ESR-spectra are obtained in the presence or absence of cysteamine (Fig. 46a). The signals obtained immediately after warming to room temperature (b) are only marginally different. Only spectra taken 6 days after warming up to room temperature (c) are significantly different: on the right hand side, the spectrum of the RS˙ radical appears (arrow). Although this example, because of the pronounced dependence on time and temperature, seems to indicate a restitution mechanism **rather** than the scavenging of hydrogen atoms (since these would be expected to react within a very short time of warming up), the results are not an unambiguous proof of a restitution process. It should be noted that the process of hydrogen donation in dry samples cannot readily be visualized, because of the restriction of diffusion of the particles involved.

b) Protection in Solutions

In the field of molecular radiation biology, a primitive protection by competition is obtained by irradiating phages in solutions containing between 1% and 5% nutrient broth. This contains high molecular weight organic compounds which react very effectively with the different radiation-induced water radicals, and protect the phages to a large extent against indirect attack. There are also a number of chemical compounds that scavenge certain water radicals *selectively*. For example, T1-phages irradiated in buffer are protected by nitrate ions, which react effectively with hydrated

86

electrons (Bachofer and Pollinger, 1954):

$$NO_3^- + e_{aq}^- + H^+ \rightarrow NO_2 + OH^-. \qquad (6.26)$$

As an example of the protective effect of a typical $OH^{.}$ scavenger, the protection of phage T1 by ortho- or para-aminobenzoic acid (PABA) can be mentioned (Bachofer and Hartwig, 1956), as well as the protection of infectious phage DNA by the addition of potassium iodide (Blok, 1967). There are practically no selective $H^{.}$ scavengers, but most radical scavengers react to some extent with atomic hydrogen; for example oxygen, which is also a good electron scavenger:

$$O_2 + e_{aq}^- \rightarrow O_2^- \qquad (6.27)$$

$$O_2 + H^{.} \rightarrow HO^{.}_2. \qquad (6.28)$$

These two equations represent the extension of the description of radiolysis of water, which is necessary if oxygen is present. If the dissolved molecules are particularly sensitive to attack by $H^{.}$ and e^-, then oxygen will exert a protective action if the $HO_2^{.}$ radical or its ionic form O_2^- are less damaging. However, the protective action observed is generally small, as most biomolecules are more sensitive to OH radicals. If, therefore, the OH radicals are scavenged by KI, then quite a large protection factor can be observed in the presence of oxygen: for example, Blok (1967) found that in the irradiation under oxygen of infectious $\Phi X\,174$-DNA (see Chapter 11.2) in solutions containing KI, the radiation sensitivity was 150 times smaller than under anaerobic conditions. The action of oxygen and many other radical scavengers in dry substances is fundamentally different from the action in solutions, as will be discussed in detail in Chapter 8. For further details of the action of radical scavengers, see the review paper by Nakken (1965).

Sulphydryl compounds are important protectors in solutions, as well as in dry systems. In many cases, these generally toxic compounds exert protection even when the addition of radical scavengers has already produced maximum competitive protection. For example, the dependence of the D_{37} for the inactivation of phage T1 on the concentration of the SH-containing protective agent cysteamine is plotted in Fig. 47. Nutrient broth itself already gives a competition protection with a factor of 7. At high concentrations, cysteamine exerts additional protection, the magnitude of which does not depend on the presence of other radical scavengers, such as broth.

This additional protection afforded by cysteamine to systems that are already at a maximum level of competitive protection, could be considered as indicating the presence of restitution mechanisms (equation (6.25)). There is not complete agreement on this, although restitution seems more likely to occur in solution than in the dry state. Table 6 shows that the reaction of equation (6.25) is energetically possible.

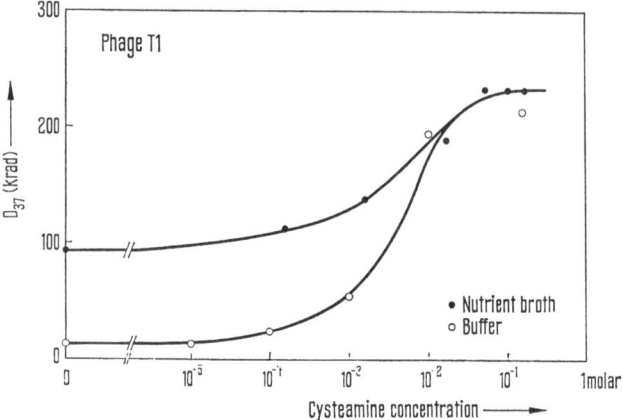

Fig. 47. 37%-dose (D_{37}) for the inactivation of T1-bacteriophages in nutrient broth and buffer solution as a function of the concentration of the protective agent cysteamine. (Hotz and Müller, 1962)

Very recently the problem of restitutive protection was investigated by means of the ESR-flow method (Nicolau and Dertinger, 1970). Chemically generated OH radicals reacted with organic target molecules to produce free radicals. When a cysteine solution was admixed with the liquid containing the free radicals immediately before passage through the ESR-cavity, the

Table 6. *Comparison of the binding energy of hydrogen in the various bonds involved in the following scheme of restitutive protection:*

$$H_2O \longrightarrow H^{\cdot} + OH^{\cdot}$$
$$OH^{\cdot} + MH \longrightarrow M^{\cdot} + H_2O$$
$$M^{\cdot} + RSH \longrightarrow RS^{\cdot} + MH$$

Bond	[eV]	Energy [kcal/mole]
HO—H	5	118
N—H, C—H	4.1—4.4	95—100
RS—H	3.7	88

spectrum resulting from these radicals disappeared and the RS˙-signal was observed. This was taken as an indication of a hydrogen donation process occurring according to equation (6.25).

An indication of the reaction of SH compounds with damaged biomolecules was provided by Blok (1967). He found that in the irradiation of aqueous solutions of ΦX 174-DNA, not only did the thioglycol-containing solutions show the expected protection against inactivation, but at the same time the mutation rate increased; at a thioglycol concentration of $1.5 \cdot 10^{-3}$ M, $2.1 \cdot 10^{-3}$ mutations per lethal event were measured, while the mutation rate in a pure solution was only just above the observable lower limit. This is not a result of the high doses necessitated by the protective action of thioglycol, since the same protection is obtained with deoxyguanylic acid but the mutation rate is a factor of 14 lower than the value obtained with thioglycol. As, in addition, the inactivation rate in the presence of thioglycol is proportional to the rate of induction of mutations, it can be concluded that the protective agent reacts with the DNA radicals. The resulting chemical changes are, however, not lethal but lead only to an increased probability of mutation. This fact can be tentatively expressed by stating that the protective reaction does not always lead to a complete restitution of the original state. This does not seem unreasonable, if the fact that the action of radiation on DNA liberates not only hydrogen, but also other groups, is taken into account. If hydrogen is substituted in such a site by the restitution mechanism, then the resultant change in the base may induce a mutation.

Further evidence of the existence of restitutive protection by sulphydryl compounds is obtained from the analysis of the oxygen effect in Chapter 8. This discussion has shown that not only is our understanding of protective and sensitizing reactions altogether inadequate, but also the closely related understanding of the mechanisms of direct and indirect action of radiation. Many systematic and well-directed experiments will be necessary before a picture free from contradictions (although possibly not clear in every detail) of the molecular processes of the action of radiation can be drawn.

References

Alexander, P., Charlesby, A.: In: Radiobiology symposium. Eds.: Z. M. Bacq and P. Alexander. London: Butterworth 1955, p. 49.
Bachofer, C. S., Hartwig, Q. L.: Radiat. Res. 5, 528 (1956).
— Pollinger, M. A.: J. gen. Physiol. 37, 663 (1954).
Bacq, Z. M., Alexander, P.: Fundamentals of radiobiology. Oxford: Pergamon Press 1961.
Berger, K. U.: Z. Naturforsch. 24 b, 722 (1969).
Bhattacharjee, S. B.: Radiat. Res. 14, 50 (1961).
Blok, J.: In: Radiation research. Ed.: G. Silini. Amsterdam: North-Holland Publ. Co. 1967, p. 423.
ten Bosch, J. J., Braams, R.: Cited by Braams (1963).
Braams, R.: Nature 200, 752 (1963).
Butler, J. A. V., Robins, A. B., Rotblat, J.: Proc. roy. Soc. 256, 1 (1960).
Buxton, G. V.: In: Radiation research. Ed.: G. Silini. Amsterdam: North-Holland Publ. Co. 1966, p. 423.
Gordy, W., Miyagawa, I.: Radiat. Res. 12, 211 (1960).
Günther, W., Jung, H.: Z. Naturforsch. 22 b, 313 (1967).
Heitkamp, D., Merwitz, O., Späth, H.: Z. Naturforsch. 23 b, 403 (1968).
Heller, H. C., Cole, T.: Proc. nat. Acad. Sci. (Wash.) 54, 1486 (1965).
Henriksen, T., Sanner, T., Pihl, A.: Radiat. Res. 18, 163 (1963).

Hotz, G., Müller, A.: Z. Naturforsch. **17 b**, 34 (1962).
Howard-Flanders, P.: Nature **186**, 485 (1960).
Hutchinson, F.: Radiat. Res. Suppl. **2**, 49 (1960).
— Presten, A., Vogel, B.: Radiat. Res. **7**, 465 (1957).
Jung, H.: Z. Naturforsch. **20 b**, 764 (1965).
— Z. Naturforsch. **21 b**, 1165 (1966).
— Kürzinger, K.: Radiat. Res. **36**, 369 (1968).
— — Z. Naturforsch. **24 b**, 328 (1969).
— Schüßler, H.: Z. Naturforsch. **21 b**, 224 (1966).
Müller, A., Dertinger, H.: Z. Naturforsch. **23 b**, 83 (1968).
Nakken, K. F.: In: Current topics in radiation research, Vol. I. Eds.: M. Ebert and A. Howard. Amsterdam: North-Holland Publ. Co. 1965, p. 49.
Nicolau, Cl., Dertinger, H.: Radiat. Res., in press (1970).
Ormerod, M. G., Alexander, P.: Nature **193**, 290 (1962).
Pauly, H., Pfister, H., Rajewsky, B.: Biophysik **3**, 36 (1966).
Phillips, G. O. (Ed.): Energy transfer in radiation processes. Amsterdam: Elsevier Publishing Company 1966.
Sommermeyer, K., Stegle, J., Schnepel, G. H.: Atompraxis **13**, 20 (1967).
Spikes, J. D., Straight, R.: Ann. Rev. Phys. Chem. **18**, 409 (1967).
Spinks, J. W. T., Woods, R. J.: An introduction to radiation chemistry. New York: John Wiley & Sons 1964.
Swallow, A. J.: Radiation chemistry of organic compounds. Oxford: Pergamon Press 1960.
Tanooka, H., Hutchinson, F.: Radiat. Res. **24**, 43 (1965).
Tobias, C. A., Brustad, T., Manney, T.: In: The initial effects of ionizing radiations on cells. Ed.: R. J. C. Harris. London, New York: Academic Press 1960, p. 257.
Zimmer, K. G., Bouman, J.: Physikal. Zschr. **45**, 298 (1944).

Chapter 7. The Temperature Effect

In Chapter 5, target molecular weights for various enzymes were calculated from the 37%-doses measured at room temperature, and the results compared with the true molecular weights of the irradiated biomolecules. As there is no rational reason to attach any special significance to the results obtained at room temperature, the dependence of inactivation rate on the temperature during exposure will now be examined. In some ways, this chapter could be considered as an extension of the target theory; furthermore, some additional features of the indirect effect (see Chapter 6) will become apparent, the significance of which is enhanced by the observation of a remarkable uniformity in the temperature-dependence of many biological systems under irradiation.

7.1. Experimental Observations

The term "temperature effect" refers to the fact that the radiation sensitivities of many macromolecules and biological systems decrease with decreasing exposure temperature, which is reflected in a reduction of the slope of the dose-response curves. Fig. 48 shows an example of this phenomenon, namely the inactivation of dry RNase by 2 MeV protons at two different

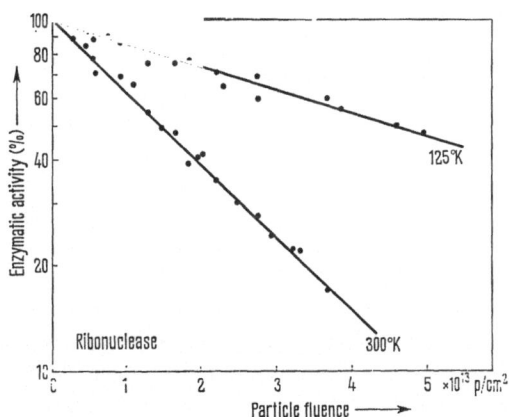

Fig. 48. Inactivation of ribonuclease by 2 MeV protons at 125 °K and 300 °K. (Günther and Jung, 1967)

temperatures. The example of inactivation of trypsin by accelerated carbon ions (Fig. 49) demonstrates that the response depends on the temperature during irradiation, at least in the inactivation of macromolecules, since changes in the temperature of the trypsin samples before or after irradiation have no influence on the radiation sensitivity determined at 300 °K.

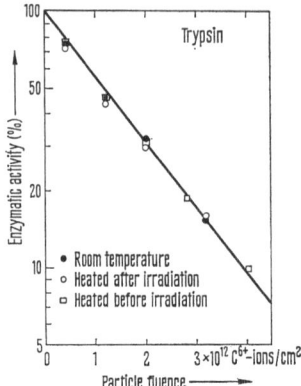

Fig. 49. Inactivation of trypsin by accelerated carbon ions at room temperature, and the effect of warming the samples to 361 °K before and after irradiation. (Brustad, 1964)

This latter observation does not, in general, apply to autonomous microorganisms, such as bacteria. A post-irradiation temperature effect is often observed, such as the influence of incubation temperature, which is due to the enzymatically controlled systems for the repair of radiation damage. This phenomenon is not, however, considered to be a manifestation of the temperature effect proper, and will therefore be excluded from this discussion.

The fact that the temperature effect is a fundamental phenomenon could be taken as a confirmation of the view that biological inactivation can be correlated with elementary lesions produced during the early physico-chemical phase of radiation action (see Fig. 2). In the case of trypsin, for example, the rate of inactivation shows the same temperature-dependence as the production of free radicals (Fig. 50). This indicates that the action of radicals is, at least in this case, an important primary step in the inactivation process.

Another type of temperature effect, already referred to in the discussion of the indirect effect (see Chapter 6.4), is the difference between the radiation sensitivities of biomolecules in solution irradiated above and below the freezing point. It will be shown later (see Fig. 56) that freezing reduces the sensitivity of an aqueous solution of trypsin by a factor of 100. This is caused by restriction of the diffusion of radiation-induced water radicals

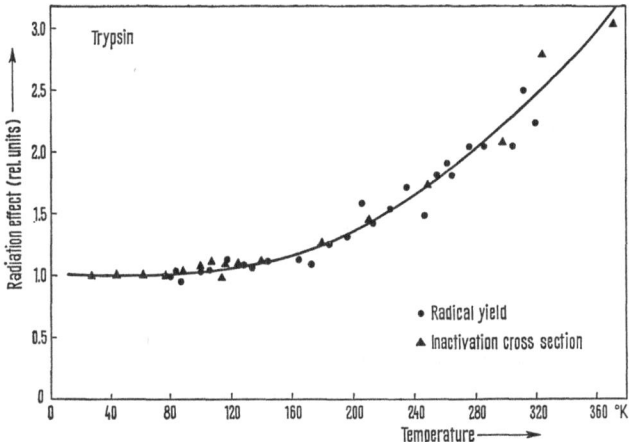

Fig. 50. Temperature dependence of inactivation and radical formation in trypsin. (Henriksen, 1966)

in ice, which in turn reduces their interaction with the enzyme molecules present. However, the contribution of free radicals does not vanish completely, even in frozen solutions, as is shown by the fact that the rate of inactivation of trypsin below the freezing-point still depends on temperature (see Fig. 56). These results suggest that a temperature effect is indicative of the presence of indirect radiation effects, and this suggestion will be examined more closely, and confirmed, in the following sections.

7.2. Temperature Effect and the Indirect Action of Radiation

Precisely formulated, this suggestion states that the temperature-dependent portion of radiation effects is caused by the attack of small radicals, which is known to occur in both wet and dry systems (see Chapter 6.4). It is, therefore, assumed that the temperature effect reflects the temperature-dependence of a rate constant k, e. g. that of the reaction of hydrogen atoms with biological molecules. The temperature-dependence is, in this case, given by the well known relationship:

$$k = k' \, e^{-E/RT}, \qquad (7.1)$$

where k' is a constant, T the absolute temperature, R the universal gas constant (1.986 cal·degree^{-1}·mole^{-1}), and E the activation energy of the reaction under discussion. The exponential factor in equation (7.1) is, for a Boltzmann energy distribution, equal to the relative number of molecules which at a temperature T possess the activation energy E. If it is assumed that the overall action of radiation consists of a direct (temperature-independent) and an indirect effect, then the following statements can be made for radiation sensitivity (for sparsely ionizing radiations), and for the inactivation cross section (for corpuscular radiations):

γ-rays: $\qquad\qquad 1/D_{37}^{\nu}=k_0+k_1\cdot e^{-E/RT}$ (7.2)

charged particles: $\qquad \sigma=\sigma_0+\sigma_1\cdot e^{-E/RT}$. (7.3)

The problem is, first of all, to investigate the validity of statements (7.2) and (7.3), and to see whether it is possible to determine the activation energy experimentally, on the basis of which conclusions could be drawn about the underlying reactions. For this purpose, equation (7.3) is rewritten as:

$$\ln(\sigma-\sigma_0)=\ln\sigma_1-E/RT .$$ (7.4)

Plotting of $\ln(\sigma-\sigma_0)$ against $1/T$ ("Arrhenius-plot") should yield a straight line of slope E/R, which immediately yields the activation energy E. If there are a number of processes with different activation energies contributing to the temperature-dependent damage, equation (7.3) can be extended correspondingly:

$$\sigma=\sigma_0+\sigma_1\cdot e^{-E_1/RT}+\sigma_2\cdot e^{-E_2/RT}+ \ldots$$ (7.5)

The extent to which the statement (7.5) is able to describe the experimental results can now be examined. Fig. 51 shows the cross section for the inactivation of RNase by 2 MeV protons as a function of reciprocal temperature. The cross section decreases continuously with increasing $1/T$, i. e. with decreasing temperature, approaching a constant value below 100 °K. In all the experiments which have so far been carried out at very low temperatures (Webb et al., 1958; Brustad, 1964; Hotz and Müller, 1968; Uenzelmann, 1968), no significant variation in radiation sensitivity has been found between 4 °K and 100 °K. If the constant fraction σ_0 is subtracted from the experimental points, the values obtained at low temperature lie on a straight line with a slope corresponding to an activation energy of 1 kcal/mole. The deviation from this line at higher temperatures indicates the superposition of another straight line with an activation energy of 6.5 kcal/mole (Fig. 51).

The example of ribonuclease is not, however, the only case where the influence of temperature can be expressed by equation (7.5): Günther and Jung (1967) analyzed many of the known experiments, and found that the temperature-dependence of radiation action can be described by three components, one of which is temperature-independent, while the other two have activation energies of 1 kcal/mole and 4—6 kcal/mole, respectively (Table 7). This relationship applies not only to the inactivation of different kinds of biological systems, but also to ESR-determinations of the yield of radiation-induced radicals in dry organic compounds.

This now leads to the problem of which particular indirect processes are associated with these activation energies. While the role of the 4—6 kcal/mole-component remains essentially unresolved, Kürzinger (1969) has

94

Table 7. *Compilation of the activation energies* E_1 *and* E_2 *for the description of the reaction cross section* σ *as a function of absolute temperature* T *by equation (7.5). In all the results quoted, a temperature-independent part* σ_0 *was found, while the term in* E_2 *could only be determined in experiments extending to temperatures higher than room temperature.* (Günther and Jung, 1967)

Object	Radiation	Test	E_1	E_2 [kcal/mole]	Authors
Glycine	330 MeV-Ar	Radicals	0.9	3.7	Henriksen, 1966
Ribonuclease	3 MeV-e	Inactivation	1.06	6.1	Fluke, 1966
Ribonuclease	2 MeV-p	Inactivation	1	6.5	Kürzinger and Jung, 1968
Ribonuclease	2 MeV-d	Inactivation	1.05	—	Günther and Jung, 1967
Lysozyme	3 MeV-e	Inactivation	0.62	2.54	Fluke, 1966
Lysozyme	33 MeV-α	Radicals	1	5	Henriksen, 1966
Lysozyme	100 MeV-C	Radicals	1	4	Henriksen, 1966
Trypsin	18 MeV-d	Inactivation	1.1	4.5	Brustad, 1964
Trypsin	33 MeV-α	Inactivation	1.2	5	Brustad, 1964
Trypsin	33 MeV-α	Radicals	0.95	3.6	Henriksen, 1966
Hyaluronidase	1 MeV-e	Inactivation	1.1	—	Vollmer and Fluke, 1967
Invertase	4 MeV-d	Inactivation	1	6	Pollard et al., 1952
Invertase	8 MeV-α	Inactivation	0.95	6	Pollard et al., 1952
ΦX 174-DNA	10 MeV-e	Inactivation	missing	6.3	Hotz and Müller, 1968
ΦX 174-Phage	2 MeV-p	Inactivation	1	—	Günther and Hermann, 1967
Phage T1	4 MeV-d	Inactivation	missing	5.4	Adams and Pollard, 1952
Phage T1	50 kVp-X	Inactivation	1.1	—	Bachofer et al., 1953
Phage T1	2 MeV-p	Inactivation	1	—	Hermann, 1966
Phage BU-T1	2 MeV-p	Inactivation	1	—	Hermann, 1966
B. megaterium spores	50 kVp-X	Inactivation	1.06	—	Webb et al., 1958

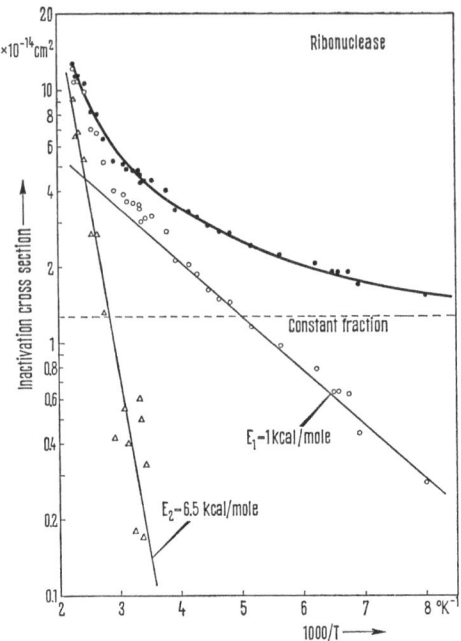

Fig. 51. Cross section for the inactivation of ribonuclease by 2 MeV protons as a function of the reciprocal of the absolute temperature (Arrhenius-diagram). See the text concerning the determination of the activation energies E_1 and E_2. (Kürzinger and Jung, 1968)

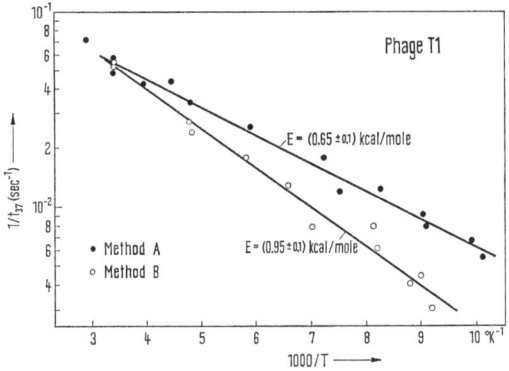

Fig. 52. Sensitivity ($1/t_{37}$; $t_{37} = 37\%$-exposure time) of T1-bacteriophages to atomic hydrogen as a function of the reciprocal of the absolute temperature (Arrhenius-diagram). The hydrogen atoms were produced by the irradiation of plastic foils with 2 MeV protons (see Figure 44). Method A: Foil at room temperature, phages at the temperature T. Method B: Foils and phages at the same temperature T. (Kürzinger, 1969)

carried out an interesting experiment which indicates that the 1 kcal/mole-component may be associated with the radiation-induced hydrogen atoms, which are released in the irradiation of organic materials (see Chapter 6.4) and may then react with undamaged molecules. Using an arrangement similar to that sketched in Fig. 44, phage T1 was exposed to thermal hydrogen atoms produced by the irradiation of a plastic foil with 2 MeV protons. The inactivation increased exponentially with the exposure time t, at all temperatures examined. Plotting $1/t_{37}$ against the reciprocal of the temperature during irradiation, two straight lines with activation energies of 0.65 and 0.95 kcal/mole respectively were obtained (Fig. 52). In the first case (Method A), the irradiated plastic foil was maintained at room temperature and the yield of hydrogen atoms therefore remained constant, while the bacteriophages were cooled to different temperatures. The second case (Method B), in which the foil and sample were maintained at the same temperature, is obviously of greater biological interest, since when macromolecules are irradiated at low temperatures, the molecules providing the hydrogen are at the same temperature as the molecules reacting with the hydrogen. The activation energy of (0.95 ± 0.1) kcal/mole obtained under these conditions, is in close agreement with the value $E_1 = 1$ kcal/mole measured in nearly all the systems quoted in Table 7.

7.3. LET-Dependence of the Temperature Effect

The results of the previous section indicate that the hypothesis of the temperature effect as presented here may quite possibly be correct. A closer examination of the consequences of this hypothesis for the concepts of target theory is therefore desirable. For this purpose, the LET-dependence of the radiation sensitivity of trypsin at different temperatures may be considered (Fig. 53). The curves presented do not take the "classical" form, i. e. they do not approach zero asymptotically at high LET. This is to be expected in any case, since the theory of Butts and Katz is applicable to the small trypsin molecule (see Chapter 5.3). The sensitivities are constant at low LET, and then change to a lower, apparently constant sensitivity level at high LET. The difference between the sensitivities at high and low LET becomes smaller with decreasing temperature, until the radiation sensitivity finally becomes independent of LET below 100 °K.

The curves in Fig. 53 can be explained to some extent by the theory of Butts and Katz (equation 5.15) since the temperature-dependence is implicit in the concept of D_{37}^x . A decrease in temperature causes an increase in D_{37}^x, i. e. a progression from top to bottom within the family of curves shown in Fig. 35. As the D_{37}^x increases, so too does the region in which σ is proportional to LET and insensitive to the influence of charge. This explains the independence of radiation sensitivity from LET (Fig. 53) for $T < 100$ °K, and also the fact that the "transition point" of the sensi-

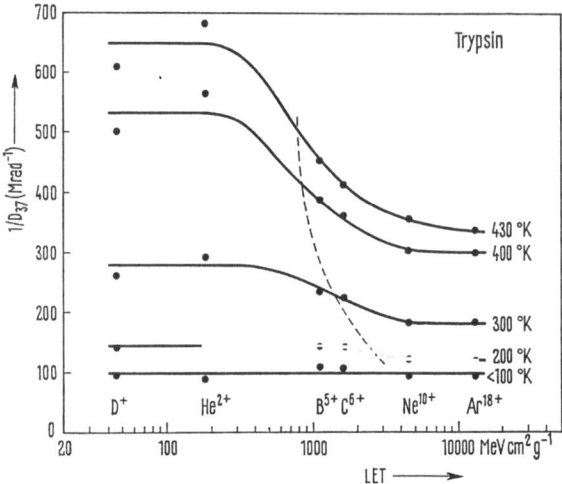

Fig. 53. Radiation sensitivity ($1/D_{37}$) of trypsin as a function of linear energy transfer (LET) at different temperatures. The broken line connects the mean values of the radiation sensitivities at low and at high LET. (Brustad, 1964)

tivity is shifted to higher LET with decreasing temperature (Fig. 53, broken line). The significance of the constant sensitivity level at high LET has not yet been explained completely. It may possibly be associated with the "thermal spike" phenomenon, which will be discussed in the next section. In any case, equation (5.15) should be capable of describing the conditions in the region of low and intermediate LET where the sensitivity is constant. Its development into a power series, neglecting higher order terms gives:

$$\sigma = A \cdot (L/\varrho)/D_{37}^{\gamma}, \qquad (7.6)$$

where the quantity A takes account of the value of the integral in equation (5.15) and is assumed to be LET-independent for a first order approximation. The quantities occuring in the exponential factor (charge and velocity) are expressed approximately by the LET. Equation (7.6), from what has been said so far, applies only with large values of D_{37} and at low and intermediate values of LET. The theory of Butts and Katz attributes the action of densely ionizing particles to the release of secondary electrons, the effectiveness of which is supposed to correspond to a randomly distributed γ-ray dose. If, therefore, the sensitivity to sparsely ionizing radiation is described by equation (7.2), as has been shown for RNase by Kürzinger and Jung (1968), then equation (7.2) can be entered into equation (7.6) giving:

$$\sigma = A \cdot (L/\varrho) \cdot (k_0 + k_1 \cdot e^{-E/RT}) = \sigma_0 + \sigma_1 \cdot e^{-E/RT}. \qquad (7.7)$$

This means that the semi-empirical statement (7.3) is justified by the theory of Butts and Katz, which in addition predicts that both σ_0 and σ_1 are directly proportional to LET. According to equation (5.8), therefore, both the constant and the temperature-dependent parts of radiation sensitivity ($1/D_{37}$) should be independent of LET. In the case of RNase, this conclusion can be confirmed directly by the following results.

3 MeV electrons (Fluke, 1966):

$$1/D_{37} = (0.005 + 0.06 \cdot e^{-1060/RT} + 25 \cdot e^{-6100/RT}) \, \text{Mrad}^{-1} \qquad (7.8)$$

2 MeV protons (Kürzinger and Jung, 1968):

$$1/D_{37} = (0.0048 + 0.06 \cdot e^{-1000/RT} + 52 \cdot e^{-6500/RT}) \, \text{Mrad}^{-1} \qquad (7.9)$$

2 MeV deuterons (Günther and Jung, 1967):

$$1/D_{37} = (0.0055 + 0.07) \cdot e^{-1050/RT}) \, \text{Mrad}^{-1} . \qquad (7.10)$$

Since the irradiation with 2 MeV deuterons was only carried out below room temperature, the second temperature-dependent component could not be measured. Its determination, in any case, contains a large inherent error, because it is obtained by twice taking the difference between values containing certain errors. Consequently, the differences between the coefficients in the third term in equation (7.8) and (7.9) are not statistically reliable. Furthermore, the fact that the significance of this second temperature-dependent component is not yet clear should not be overlooked; it may have genuine dependence on LET. Equation (7.7), therefore, proves to be a basic relationship, by which, within certain limits, the target theory can be extended to cover temperature-dependent processes, in so far as these can be attributed to the indirect actions of radiation.

7.4. The "Thermal Spike" Model

In conclusion, the question of whether or not there are other possibilities of explaining the temperature effect must be asked. The observation that at high LET the radiation sensitivity apparently reaches a constant level is used as the starting point for this discussion. This may possibly reflect certain "track effects", which increase in importance as the LET increases, so that a consideration of the distribution of δ-rays may not be adequate to give an exact description of the inactivation cross section in this case. The mechanism of such "track effects" could be based on the fact that, due to the high rate of energy deposition along the track of a densely ionizing particle, a local high temperature can be generated, and that biological objects within this cylindrical heat-column (spike) of a certain diameter may, therefore, be thermally inactivated. This is quite an old idea, going back to Dessauer (1923). It was, however, disregarded until, in more recent times, the idea of "point heat" was resurrected by Norman and colleagues (Norman and Spiegler, 1962; Ingalls et al., 1964) and extended to what is now known as the "thermal spike model".

For the quantitative formulation of this problem a particle track is considered as a linear burst of heat, of a magnitude determined by the LET, from which a wave of heat expands radially by conduction. This may produce a damaging temperature rise within a cylindrical region. The mathematical formalism is provided by the heat conduction equation, the solution of which under these conditions is:

$$T(r, t) = T(0, 0) + \frac{L/\varrho}{4\pi D t c} e^{-\frac{r^2}{4Dt}}, \qquad (7.11)$$

where t is the time after the passage of a particle, r the radial distance from the particle track, D the thermal diffusion coefficient, c the specific heat, and T the absolute temperature. The "lifetime" of the exponential function (7.11) is given by

$$t_0 = \frac{r^2}{4D}. \qquad (7.12)$$

Considering the rapid variation of this exponential function, equation (7.11) can to a good approximation be written in the form:

$$t < t_0 : T(r, t) = T(0,0)$$
$$t > t_0 : T(r, t) = T(0,0) + \frac{L/\varrho}{4\pi D t c}. \qquad (7.13)$$

In order to be able to determine the cross section of the thermal spike within which thermal inactivation will occur, it is necessary to postulate that $T(r, t)$, on the left-hand side of the equation (7.11), achieves a certain decomposition temperature T_D. $T(0,0)$ is then equal to the irradiation temperature T, while the quantity of $4\pi D t$ in the denominator of equation (7.13) represents the cross section over which the heat wave has increased the temperature above the value T_D at a time t. The inactivation cross section is then given by:

$$\sigma = \frac{L/\varrho}{c(T_D - T)}. \qquad (7.14)$$

Of course, the time t in equation (7.13) must not be chosen too small, but rather, the temperature T_D must be maintained for a certain minimum period. As the "lifetime" for the thermal decomposition of macromolecules is of the order of 10^{-9} seconds, the validity of equation (7.14) is restricted to LET values above 10^3 MeV cm^{2}g^{-1}. The choice of decomposition temperature and specific heat causes some headache. Norman and Spiegler (1962) obtained quite a good description of the LET-dependence for the inactivation cross sections of enzymes, by assuming T_D to be the mean decomposition temperature of amino acids (approximately 300 °C) and c to be in the range 0.3 to 0.4 cal·g^{-1}·degree^{-1}.

It is difficult to decide whether or not the thermal spike inactivation model has any real significance. It may be necessary to divide the LET in Fig. 53 into two regions, one above and one below about 1000 MeV cm^2 g^{-1}, with the theory of Butts and Katz applying below, and the thermal spike

100

model applying above this LET value, each providing the required LET-proportionality of the inactivation cross section within its limits of validity. The overall temperature-dependence of the cross section can, therefore, be described by:

$$\sigma = \sigma_\delta(L, T) + \frac{L/\varrho}{c\,(T_D - T)}, \qquad (7.15)$$

where $\sigma_\delta(L,T)$ gives the inactivation cross section according to equation (7.7).

The validity of equation (7.15) cannot be tested rigorously with the experimental results as yet available. The results shown in Fig. 34 do indicate, however, that at least the LET-dependence can be described accurately by equation (7.15), because after an initial linear rise due to the first term of equation (7.15) and a later decrease in slope, the slope of the three curves increases once again. This may indicate the influence of the second term of equation (7.15).

This discussion has shown that the influence of temperature on the sensitivity of irradiated biomolecules is still far from being understood in detail. Nevertheless, the investigations of the last few years, and the ideas which have arisen about possible reaction mechanisms, do provide starting points for further *directed* investigations into this problem. There is, therefore, some hope that this important parameter which modifies radiation sensitivity will be understood in the not-too-distant future, so that further interesting insights into the mechanisms of the action of radiation may be obtained.

References

Adams, W. R., Pollard, E. C.: Arch. Biochem. Biophys. 36, 311 (1952).
Bachofer, C. S., Ehret, C. F., Mayer, S., Powers, E. L.: Proc. nat. Acad. Sci. (Wash.) 39, 744 (1953).
Brustad, T.: In: Biological effects of neutron and proton irradiations, Vol. II. Vienna: Internat. Atomic Energy Agency 1964, p. 404.
Dessauer, F.: Z. Physik 12, 38 (1923).
Fluke, D. J.: Radiat. Res. 28, 677 (1966).
Günther, H. H., Hermann, K. O.: Z. Naturforschg. 22 b, 53 (1967).
Günther, W., Jung, H.: Z. Naturforschg. 22 b, 313 (1967).
Henriksen, T.: Radiat. Res. 27, 694 (1966).
Hermann, K. O.: Z. Naturforschg. 21 b, 678 (1966).
Hotz, G., Müller, A.: Proc. nat. Acad. Sci. (Wash.) 60, 251 (1968).
Ingalls, R. B., Spiegler, P., Norman, A.: J. Chem. Phys. 41, 837 (1964).
Kürzinger, K.: Int. J. Radiat. Biol. 16, 1 (1969).
— Jung, A.: Z. Naturforschg. 23 b, 949 (1968).
Norman, A., Spiegler, P.: J. appl. Phys. 32, 2658 (1962).
Pollard, E. C., Powell, W. F., Reaume, S. H.: Proc. nat. Acad. Sci. (Wash.) 38, 173 (1952).
Uenzelmann, J.: Dissertation, University of Heidelberg, 1968.
Vollmer, R. T., Fluke, D. J.: Radiat. Res. 31, 867 (1967).
Webb, R. B., Ehret, C. F., Powers, E. L.: Experientia 14, 324 (1958).

Chapter 8. The Oxygen Effect

The term "oxygen effect" refers to the observation that the radiation sensitivity of macromolecules and biological systems irradiated in the presence of oxygen or air is generally higher than when they are irradiated under vacuum or in an inert atmosphere. This only applies, however, with ionizing radiations; in UV irradiation experiments, an oxygen effect is only rarely observed. As with the temperature effect, justice is not done to the oxygen effect by treating it merely as a troublesome side-effect of radiation action. It is actually a phenomenon of great heuristic importance for the elucidation of the molecular nature of radiation damage. It is a pity that here, as in many other aspects of radiation biology, relevant experiments are scarce and the many facets of the oxygen effect tend in general to produce confusion rather than understanding. It is therefore not surprising that there is no satisfactory interpretation of the oxygen effect as yet. Nevertheless, an attempt will be made to describe the oxygen effect quantitatively, with the aid of known physico-chemical data and taking specific aspects of the inactivation of microorganisms into account. The chemical mechanisms underlying the oxygen effect will be studied, in the light of experiments on the radiation inactivation of biological macromolecules.

8.1. The Oxygen Effect in Macromolecules

a) Experimental Results

The macromolecules referred to in this context are enzymes and nucleic acids, whereby the most extensive and unambiguous experimental results have been obtained from experiments on the inactivation of enzymes. An oxygen effect is almost always observed in the irradiation of dry enzymes: for example, Fig. 54 shows dose-response curves for the inactivation of RNase under aerobic (O_2) and anaerobic (vacuum) conditions. The values of the D_{37} are 20 and 42 Mrad, respectively, from which the quotient of aerobic to anaerobic sensitivity, known as the *oxygen enhancement ratio*, of 2.1 is derived. In contrast, very little or no enhancement is obtained if macromolecules are irradiated in dilute aqueous solution, i. e. under conditions where indirect action predominates; protective effects may even be

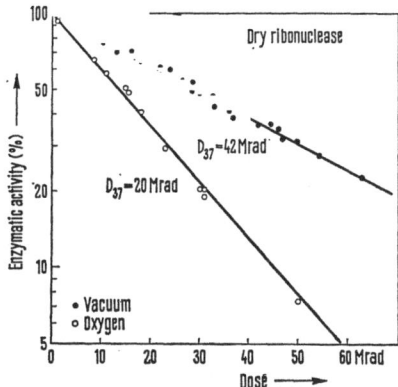

Fig. 54. Inactivation of dry ribonuclease by ^{60}Co γ-radiation in vacuo and in the presence of oxygen. (Günther and Jung, 1967)

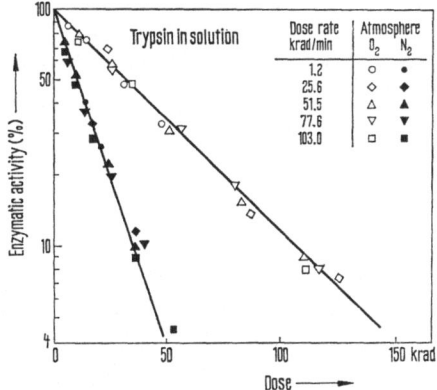

Fig. 55. Inactivation of trypsin in aerobic and anaerobic aqueous solutions (0.1 mg/ml) by 45 kVp X-rays at various dose rates. (Oksmo and Brustad, 1968)

observed (a list of examples has been given by Brustad, 1966). This last observation is illustrated by Fig. 55, which shows the dose-response curves for the radiation inactivation of trypsin in aerobic and anaerobic aqueous solutions. The radiation sensitivity obtained under a nitrogen atmosphere is 3 times as high as that obtained under oxygen, regardless of the intensity of the X-irradiation.

A further example of the protective action of oxygen in indirect systems, which is at the same time a comparison of direct and indirect action, is presented in Fig. 56. Trypsin was irradiated in the dry state and in aqueous solution under both aerobic and anaerobic conditions, and the radiation sensitivity plotted against the temperature during irradiation. In the dry state, an oxygen enhancement ratio of about 1.5 was found over the whole

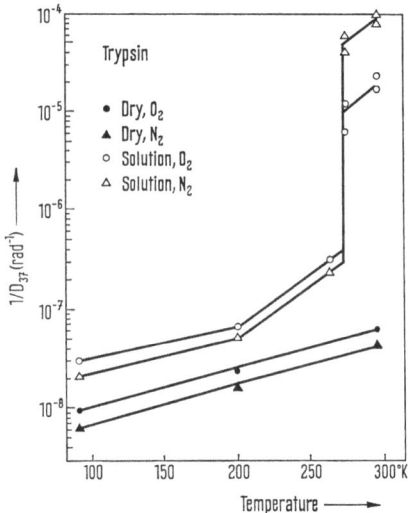

Fig. 56. Radiation sensitivity ($1/D_{37}$) of dried and dissolved trypsin under nitrogen and oxygen atmospheres as a function of irradiation temperature. (Oksmo and Brustad, 1968)

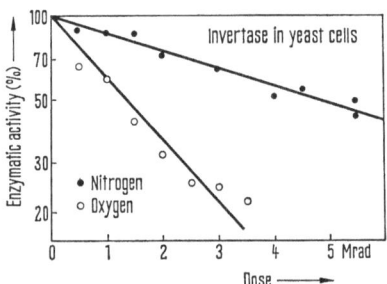

Fig. 57. Inhibition of the activity of invertase by irradiation in yeast cells with 1 MeV electrons under nitrogen and oxygen. (Hutchinson, 1961)

temperature range, while in the indirect system this was observed only at temperatures below the freezing-point. At temperatures above 0° C, the radiation sensitivity increases suddenly, and the oxygen gives significant protection.

The enhancement ratio observed for dry enzymes usually lies between 1.5 and 2, while in aqueous systems it is either small, or no enhancement occurs, or alternatively the oxygen may even protect. This may seem surprising initially, since enzymes irradiated in living cells show an oxygen enhancement factor of 3 or more (Fig. 57); whether or not this result is related to the metabolic processes of the cells can be examined by carrying out the same experiment using homogenized cells. This, however, gives the

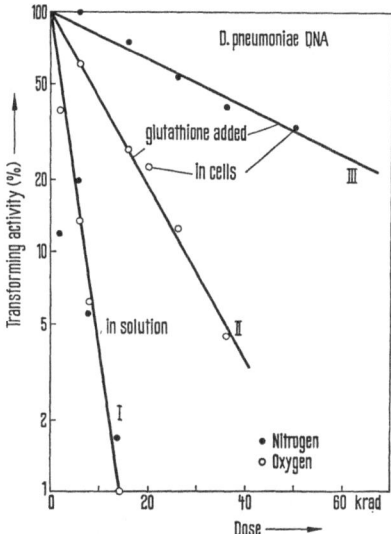

Fig. 58. Inactivation of transforming DNA from Diplococcus pneumoniae. Curve I: Irradiation in aqueous solution (0.6 mg/ml) under oxygen and nitrogen. Curves II and III: Irradiation of the DNA in vegetative cells (experimental points) and in a $1.4 \cdot 10^{-3}$ M solution of glutathione (full lines) under nitrogen and oxygen atmospheres. (Hutchinson, 1961)

same result. Therefore, there must be certain substances present in cells that are responsible for the occurence of oxygen enhancement.

This situation is not restricted to enzymes alone. Transforming DNA also shows no oxygen effect in aqueous solution (Fig. 58), while DNA irradiated in cells has a sensitivity 3.7 times greater in the presence of oxygen than under anaerobic conditions. It is interesting to note that both the sensitivity observed in cells and the oxygen effect can be reproduced *in vitro* by the suitable choice of concentration of a protector (e.g. $1.4 \cdot 10^{-3}$ molar gluthathione). One reason for the enhancing effect of oxygen in cells may therefore be the presence of substances which, like glutathione, act as protective agents. Although the list of experiments could be considerably enlarged, the experiments described are typical and display all the essential features. They show that the oxygen effect is not a clear-cut phenomenon, but that a variety of chemical reactions of the oxygen may be involved in the occurence of this effect. This will now be considered in more detail.

b) Chemistry of the Oxygen Effect in Macromolecules

The paramagnetic properties of the oxygen molecule provide the basis for an explanation of the modifying action of oxygen, as they give it a high affinity for radiation-induced radicals, with which it reacts leading to the formation of peroxy-radicals:

$$M^{\cdot} + O_2 \rightarrow MO_2^{\cdot} . \tag{8.1}$$

105

This property of oxygen does not depend on whether M' is a radical group in an irradiated biomolecule or a water radical. In addition, allowance must be made for the ability of electronegative oxygen to scavenge electrons by the mechanism of equation (6.27). This latter aspect is significant for the understanding of the oxygen effect in dry as well as in aqueous systems.

In dry systems the enhancing action of oxygen can be explained by assuming that some of the initially damaged biomolecules are restituted in the absence of oxygen. (In the next chapter some results supporting this assumption will be described). Restitution of an ionized molecule MH^+ could, for example, occur by charge neutralization with a free electron:

$$MH^+ + e^- \rightarrow MH \tag{8.2}$$

and restitution of a radical M', for example, by reaction with a hydrogen atom:

$$M' + H' \rightarrow MH . \tag{8.3}$$

In the presence of oxygen, both restitution processes are inhibited, process (8.2) by scavenging of the electron by oxygen via reaction (6.27), and process (8.3) by peroxydation of the radical M' via reaction (8.1), leading to irreversible damage.

The reverse applies in the irradiation of *aqueous systems,* where the oxygen reacts predominantly with the hydrogen radicals and the hydrated electrons. These reactions have already been encountered in the discussion of radical scavenging in Chapter 6.5, and are described by equations (6.27) and (6.28). If H' and e^-_{aq} are important in a particular inactivation process, then oxygen may protect via these two reactions. The example given in Fig. 55 shows this clearly. In cases where the damage is caused predominantly by OH radicals, oxygen usually has little influence on the radiation sensitivity of the dissolved macromolecules. If, however, a small enhancement is observed (e. g. RNase in solution has an enhancement ratio of 1.2; Jung and Schüssler, 1966), then this could be due to the scavenging of H radicals by oxygen according to reaction (6.28). The scavenging of hydrogen radicals by oxygen thereby reduces the recombination $H' + OH' \rightarrow H_2O$, and consequently increases the number of OH radicals slightly. It is therefore important to make an accurate distinction between the different modes of action of oxygen in the dry state and in dilute aqueous solutions. If the oxygen competes with restitution mechanisms for a potential lesion in a macromolecule, an enhancement might be expected as is observed, for example, in irradiated enzymes. If, however, the radical-scavenging role of oxygen predominates, then a protective action is obtained. This immediately gives a better understanding of the initially puzzling observations on the influence of oxygen on the irradiation of enzymes and DNA (Figs. 57 and 58). SH-containing substances are, as is well known, not merely good radical scavengers but can probably also restitute a damaged macromolecule, as has been discussed in connection with the hydrogen donation hypothesis (equation 6.25). If such a substance (e. g. glutathione) is added to a DNA solution, an extraordinary decrease in radiation sensitivity is obtained under anaerobic conditions (Fig. 58, curve III), which probably involves both mechanisms. In aerobic conditions, the oxygen competes with this restorative ability of glutathione, and an enhancement of the sensitivity by a factor of 3.7 is obtained (Fig. 58; curves II and III). The level of sensitivity of the unprotected solution (curve I) is not reached; this is mainly due to the fact that glutathione also acts as a radical scavenger.

In none of the examples mentioned, in which an enhancement of radiation sensitivity is observed in the presence of oxygen, is there a genuine sensitization, i. e. an increase in the number of primary lesions. The enhancing action of oxygen is always due to the reduction or cessation of restitution processes. Since the rate of restitution is obviously lower in dry systems than in solutions containing a protective agent (or than in cells), the enhancement factor is also expected to be lower. Similarly, no enhancement by oxygen is observed in dilute aqueous solutions in the absence of protective agents, where there is probably no significant restitution of primary lesions.

This concludes the discussion of the molecular aspects of the oxygen effect, which has yielded the important result that oxygen enhancement in macromolecules is always linked with the occurrence of restitution processes. This idea is of fundamental importance, and is also applicable to the inactivation of microorganisms, in which the macromolecule DNA is the primary radiation sensitive structure.

8.2. An Oxygen Effect Hypothesis

The principle behind the following explanation of the oxygen effect was first stated by Howard-Flanders (1958). It is chiefly based on formal target-theoretical considerations, taking into account, however, the idea that the oxygen effect reflects competition between restitution processes and the irreversible "peroxidation" of a primary lesion (Chapter 8.1). A remarkable feature of this hypothesis is that it can be generalized to apply to the inactivation of microorganisms, thus providing information about the nature of the basic molecular lesions as will be shown in Chapter 8.3.

The following observation is the starting point for the hypothesis: if the relative radiation sensitivity S_r (i. e. the sensitivity in the presence of oxygen divided by the sensitivity in an inert atmosphere such as N_2) is plotted as a function of oxygen concentration $[O_2]$, then systems uniformly supplied with oxygen give the characteristic curves shown in Fig. 59, which approach an upper limit at high concentrations (partial pressures).

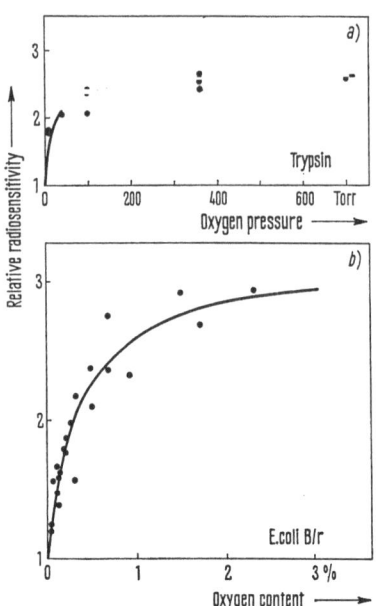

Fig. 59. a. Relative radiation sensitivity of dried trypsin to ^{60}Co γ-radiation as a function of the partial pressure of oxygen. The curve fitted through these experimental points is described by the expression $S_r = (2.6[O_2]+15)/([O_2]+15)$. (Hutchinson and Watts, 1961). b. Relative radiation sensitivity of E. coli B/r as a function of the oxygen content of the gas atmosphere. (Howard-Flanders and Alper, 1957)

They are approximately described by:

$$S_r = \frac{1/D_{37}(O_2)}{1/D_{37}(N_2)} = \frac{m_0[O_2] + k}{[O_2] + k}, \qquad (8.4)$$

where k is a constant and m_0 the maximum sensitivity obtained at high oxygen concentration. This equation, also known as the Alper formula (Alper, 1956), not only represents a formal mathematical description, but can also be derived rigorously using a suitable reaction kinetic for the oxygen effect. As this discussion is restricted to single-hit processes, the dose-response curves can be written in the following form:

$$\text{Nitrogen:} \quad N/N_0 = e^{-D/D_{37}(N_2)} = e^{-S(N_2)D} \qquad (8.5)$$

$$\text{Oxygen:} \quad N/N_0 = e^{-D/D_{37}(O_2)} = e^{-S(O_2)D}. \qquad (8.6)$$

Since an average of one damaging event has occured per object at a dose $D = D_{37}$, the quantity $S = 1/D_{37}$ is a measure of the ratio of inactivating to total lesions.

a) Macromolecules

In order to derive expression (8.4) it is necessary to assume that there are two different types of lesions, namely a potentially lethal lesion, type 1, which is only converted to lethal damage by the action of oxygen, and a type 2 lesion that is always lethal, the formation of which does not depend on the presence of oxygen. The numbers of these two types should have a definite ratio:

$$n_1/n_2 = m - 1. \tag{8.7}$$

The total number of lesions formed is then $n_1 + n_2$, and the number of lethal lesions is n_2 in nitrogen and $pn_1 + n_2$ in oxygen, where p is the probability of the conversion of type 1 damage to lethal damage by oxygen. Therefore:

$$S(N_2) = \frac{n_2}{n_1 + n_2} = \frac{1}{m} \tag{8.8}$$

$$S(O_2) = \frac{pn_1 + n_2}{n_1 + n_2} = \frac{p(m-1) + 1}{m} \tag{8.9}$$

and consequently:

$$S_r = \frac{S(O_2)}{S(N_2)} = 1 + p(m - 1). \tag{8.10}$$

To determine the probability p, and to verify the idea of the kinetics of the oxygen effect, as a competition between oxygen and a restitution process for the type 1 lesion, it is postulated:

$$\text{Type 1 lesion} \left\langle \begin{array}{l} \xrightarrow{\quad k_1 \quad} \text{repaired} \\ \xrightarrow{\quad k_2 [O_2] \quad} \text{unrepaired} \end{array} \right.$$

According to the definition of the probability, the following relationship is obtained:

$$p = \frac{k_2 [O_2]}{k_2 [O_2] + k_1} = \frac{[O_2]}{[O_2] + k} \tag{8.11}$$

where $k = k_1/k_2$ represents the ratio of the rate constants of the model given above. With increasing O_2-concentration, p approaches 1. If the equation (8.11) is introduced into equation (8.10), then the Alper formula (8.4) with $m_0 = m$ is obtained.

b) Microorganisms

The hypothesis can now be applied to the inactivation of microorganisms, particularly bacteria. Although a detailed discussion will be given in Chapters 12 and 13, it is better to discuss the aspect of oxygen enhancement separately. If the assumption is made that the inactivation is due in most cases to a lesion in the DNA, then it is tempting to include the concept of enzymatically regulated repair of DNA in the hypothesis. The type 1 lesion, which will thus become a reparable lesion of DNA, seems more likely to be repaired by intracellular processes than the type 2 lesion which is defined as being irreversible. In the generalization of the hypothesis to include reparable systems, the assumption that every type 1 lesion that does not react with oxygen is repaired must be modified. It is postulated that in an anaerobic irradiation, any type 1 lesions that have not been repaired, together with the type 2 lesions, will contribute to the inactivation. In the presence of oxygen, the fraction of type 1 lesions that has not reacted with oxygen, but still remains unrepaired, is effective as well as all type 2 lesions and the irreversibly peroxidized type 1 lesions. If the probability that the type 1 lesion will remain unrepaired is given by u, then the equations (8.8), (8.9), and (8.10) are replaced by:

$$S(N_2) = \frac{u\,n_1 + n_2}{n_1 + n_2} = \frac{u\,(m-1) + 1}{m} \tag{8.12}$$

$$S(O_2) = \frac{u\,(1-p)\,n_1 + p\,n_1 + n_2}{n_1 + n_2} = \frac{u\,(1-p)\,(m-1) + p\,(m-1) + 1}{m} \tag{8.13}$$

$$S_r = \frac{S(O_2)}{S(N_2)} = 1 + p\,(m-1)\,\frac{(1-u)}{u\,(m-1)+1} \,. \tag{8.14}$$

If p in equation (8.14) is expressed in terms of equation (8.11), the Alper formula is once again obtained. However, the maximum sensitization is now given by:

$$m_0 = \frac{m}{u\,(m-1)+1} \,. \tag{8.15}$$

This generalization, of course, has to be regarded as a very rough approximation to the solution of the real problem. A more refined hypothesis should also allow for some repair of type 1 lesions produced under aerobic conditions. However, as will be shown in the following section the simple hypothesis derived is able to explain many of the experimental results obtained in studies on the oxygen effect.

8.3. The Oxygen Effect in Bacteria

The question of the maximum sensitization obtainable, i. e. the quantity m_0 in the Alper formula, is of special interest. The bacteria irradiated in air (corresponding to 21% O_2) have reached their largest possible oxygen enhancement, as can be seen in Fig. 59. According to equation (8.15), the enhancement increases with increasing capacity for repair ($u \rightarrow 0$), approaching the upper limit m. Such a response is actually observed in the inactivation of bacteria. Some anticipatory comments at this point, concerning the molecular nature of the two types of lesions, will give a better understanding of this phenomenon. There are numerous indications that, at least in microorganisms, the lethal events are structural changes in DNA. The extensive, irreparable type 2 lesions can then be identified with a break in both strands of the DNA helix. It is more difficult to identify type 1 lesions, which in any case are probably not a single type of lesion, but a mixture of single strand breaks in the DNA helix, damage to the bases, local denaturation, etc. (see Chapter 11).

Enzymatic repair of single strand breaks and of base damage is known to occur in reactivation processes, and is controlled by certain genes (see Chapter 13.6). In bacteria, there is a remarkable correlation between the capability for host cell reactivation (hcr) and the oxygen enhancement as expressed by equation (8.15), i. e. the hcr^+ mutants show greater sensitization than the hcr^- mutants (Table 8). The oxygen effect thus fits neatly into the picture of bacterial inactivation. However, the oxygen effect also depends on the growth and incubation conditions (cf. Alper, 1961), as these may influence the enzymatic repair processes.

Table 8. *Oxygen effect in bacteria: maximum enhancement ratio m_0 of bacterial mutants with differing capacities for host cell reactivation (hcr).* (Alper, 1967)

hcr$^+$ strain	m_0	m_0	hcr$^-$ strain
E. coli B	2.7	1.7	B_{8-1}
E. coli K12S	3.0	1.7	K12S,hcr$^-$
E. coli C	3.1	2.0	CC_4(syn$^-$)
E. coli B/r, WP2	3.0	1.7	hcr$^-$
B. subtilis BS$_{15}$	3.1	1.8	SMBL$_4$
		2.3	SMBL$_5$
P. aeruginosa 1C	3.1	2.4	HCR$_5$
		2.1	HCR$_{13}$

In this context, it should be mentioned that bacteriophages also show an oxygen effect, which depends on the repair capacity of the host cell in the manner just described. The sensitization in this case is very small, for reasons that are not clear (Ikenaga, 1968). However, a pronounced oxygen effect is obtained in the intracellular irradiation of phages (Howard-Flanders and Jockey, 1960).

8.4. Oxygen Effect and LET

Another possible way to examine the oxygen hypothesis lies in the investigation of LET-dependence. This possibility can be explained by the following "target-theoretical" argument: the type 1 lesions are "light" damage, this implying that only a small, i. e. frequently deposited, amount of energy is required for their production. If the radiation sensitivity ($1/D_{37}$) is plotted against the LET, then according to the target theory, curves without a maximum should be obtained for this type of damage. If irreparable "heavy" lesions of type 2 are associated with the rare occurrence of the deposition of large amounts of energy, then the analogous plot should lead to curves having a maximum. As the overall damage consists of both

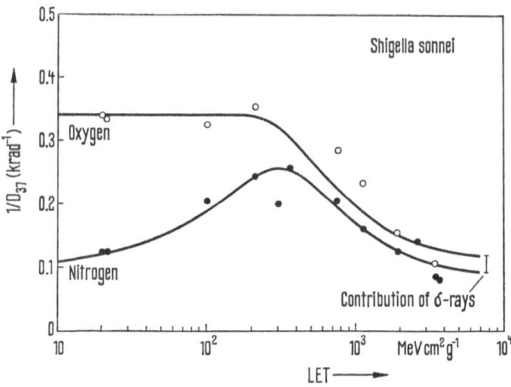

Fig. 60. Radiation sensitivity ($1/D_{37}$) of shigella sonnei as a function of the linear energy transfer (LET) under aerobic and anaerobic conditions. (Brustad, 1961)

type 1 and type 2, superposition curves should generally be obtained, with the tendency for the curves to have a maximum becoming less pronounced as the relative importance of type 1 lesions increases. This is actually found with, for example, sensitive bacteria ($u \to 1$) and for irradiations in the presence of oxygen. This is confirmed by two examples: in Fig. 60, the radiation sensitivity of bacillus Shigella sonnei under aerobic and anaerobic conditions is plotted as a function of LET. As expected from the above considerations, the anaerobic curve passes through a maximum, but not the curve obtained in the presence of oxygen. Furthermore, the figure shows that the oxygen enhancement decreases with increasing LET, since the frequency of the heavy, type 2 lesions increases with higher local energy deposition. The production of type 1 lesions is then caused preferentially by the presence of delta-rays of suitable energy, the existence of which necessarily prevents the oxygen enhancement from vanishing at high LET. This LET-dependence is correctly described by the hypothesis, since high LET is obviously identical with the condition that $m - 1 \to 0$. Consequently, $S(N_2)$ of equation (8.12) approaches $S(O_2)$ of equation (8.13). As, furthermore, the probability u of unrepaired lesions always appears multiplied by $m - 1$, the repair capacity becomes unimportant at high LET ($m - 1 \to 0$); i. e. at high LET, there is no longer any sensitivity difference between resistant and sensitive mutants. This is also confirmed by Fig. 61, which shows the radiation sensitivity of different mutant strains of E. coli as a function of LET. As expected, the maximum vanishes with decreasing capacity for repair ($u \to 1$), and in addition the differences between the radiation sensitivities of the different mutant strains decrease with increasing LET.

These two examples show that the hypothesis even in its simplified form not only describes the oxygen effect correctly, but also the sensitivity

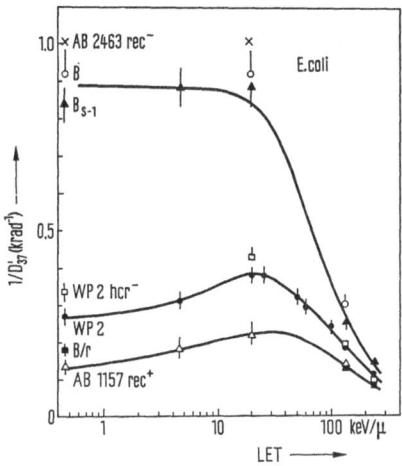

Fig. 61. Radiation sensitivity $(1/D'_{37})$ of different E. coli mutants as a function of the linear energy transfer (LET) of the radiation used. (Munson *et al.*, 1967). See Chapter 13.2 for the definition of the dose D'_{37}

of bacteria as a function of LET. In spite of the apparently good agreement found in the application of the oxygen hypothesis to this series of experimental observations, it must be emphasized once more that the basic divisions into type 1 and type 2 lesions, which initially was purely heuristic, is a gross over-simplification. Besides the results presented in Chapter 10 a series of experiments carried out by Powers and colleagues have to be mentioned. These authors irradiated spores of Bacillus megaterium in a variety of gaseous conditions, and then transferred them to a different gaseous environment for a certain period of time, before bringing the spores into contact with air for the biological test. They derived, from the corresponding modification of the inactivation rate, a "radiation sensitivity profile" (Powers and Kaleta, 1960) which shows that there are different types of oxygen dependent lesions, some of which are long-lived radicals which by a change of environment after irradiation may cause an "after effect". It is therefore clear that the assumptions made concerning the lesions responsible for the oxygen effect are over-simplified, and this has to be remembered when applying the hypothesis presented in its basic form.

References

Alper, T.: Radiat. Res. **5**, 573 (1956).
— Int. J. Radiat. Biol. **3**, 369 (1961).
— Mutation Res. **4**, 15 (1967).
Brustad, T.: Radiat. Res. **15**, 139 (1961).
— Radiat. Res. **27**, 456 (1966).
Günther, W., Jung, H.: Z. Naturforsch. **22 b**, 313 (1967).

Howard-Flanders, P.: In: Advances in biological and medical physics, Vol. VI. Eds.: C. A. Tobias and J. H. Lawrence. New York: Academic Press 1958, p. 533.
— Alper, T.: Radiat. Res. **7,** 518 (1957).
— Jockey, P.: Int. J. Radiat. Biol. **2,** 361 (1960).
Hutchinson, F.: Radiat. Res. **14,** 721 (1961).
— Watts, E.: Radiat. Res. **14,** 803 (1961).
Ikenaga, M.: Radiat. Res. **34,** 421 (1968).
Jung, H., Schüssler, H.: Z. Naturforsch. **21 b,** 224 (1966).
Munson, R. J., Neary, G. J., Bridges, B. A., Preston, R. J.: Int. J. Radiat. Biol. **13,** 205 (1967).
Oksmo, O., Brustad, T.: Z. Naturforsch. **23 b,** 962 (1968).
Powers, E. L., Kaleta, B. F.: Science **132,** 959 (1960).

Chapter 9. The Action of Radiation on Enzymes: The Example of Ribonuclease

The aim of the previous chapters has been to emphasize the basic features as well as the general "laws" of the action of radiation. This forms a sound basis for the understanding of the "biomolecular" portion of this treatise. The study begins with radiation effects on enzymes, although this should not be taken to imply that these phenomena are particularly simple. The structure of enzymes is in some ways much more complicated than that of nucleic acids, and their three-dimensional structure reflects the highly specific catalytic properties. Interest in the action of radiation on enzymes results from the fact that they are essential for the maintenance of vital processes. The aim of this chapter is to derive an approximate picture of the action of radiation on enzymes from the multitude of different observations; but the discussion will be mainly confined to a particularly well-examined enzyme, ribonuclease (RNase), since the quantity and variety of experimental data involved does not allow a collective treatment of the various enzymes.

9.1. Structure and Function of Ribonuclease

The amino acid sequence of ribonuclease (RNase) has been known since 1959. It has a molecular weight of 13,680 and consists of 124 amino acids arranged in a single chain (Fig. 62). This chain is held in a compact spatial array by four disulphide bridges. Kartha et al. (1967) succeeded in determining the spatial structure, known as the conformation. This is maintained not only by disulphide bridges, but also by hydrogen bonds between amino and carboxyl groups, electrostatic forces and hydrophobic bonds. The amino acids having non-polar side chains with little affinity for water are involved in the formation of hydrophobic bonds. In their spatial configuration, they tend to move as far as possible to the inside of the molecule. When a number of these groups come into close contact, then they form hydrophobic bonds which increase the stability of the conformation.

The sequence of individual amino acids determines the conformation of the molecules in an essentially unambiguous way. This is shown, for example, by the fact that an enzyme can be denatured by thermal or chemical means, which cause the unfolding of the molecule. In the case of

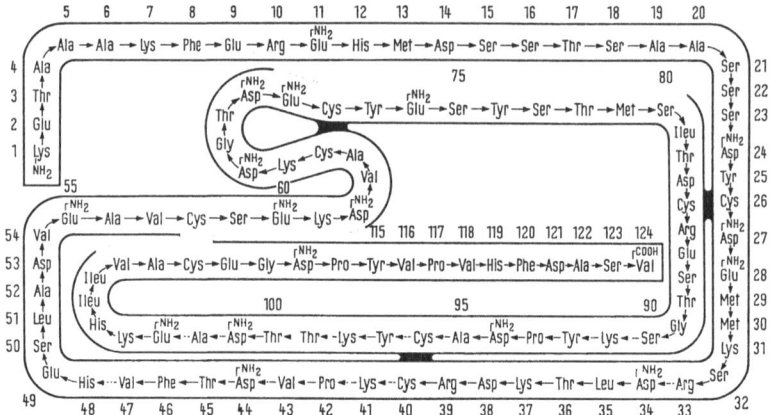

Fig. 62. Primary structure of ribonuclease with disulphide bridges. (Smyth *et al.*, 1963)

RNase, this denaturation is reversible, even when the disulphide bonds are reduced. The molecule is then an enzymatically inactive polypeptide containing 8 SH groups and with no tertiary structure; but in the presence of air it can spontaneously reoxidize to its original conformation.

RNase derives its name from the ability to degrade RNA. This decomposition occurs in two steps. The first consists of a splitting of the phosphodiester bond, and the transfer of the ester bond to the 2'-hydroxyl group of the ribose to form a cyclic diester. In the second step this pyrimidine-2',3'-cyclic phosphate bond is hydrolyzed to nucleotide-3'-phosphate. Using RNA as substrate the total reaction, and with cytidine-2',3'-cyclic phosphate the second reaction alone, can be investigated. When reference is made to enzyme activity, this will refer to the total reaction, with RNA as substrate.

9.2. Inactivation Kinetics

RNase, like most other enzymes, gives exponential dose-response curves regardless of whether the irradiation takes place in the dry state or in solution (Figs. 54 and 63), or whether the inactivation is caused by atomic hydrogen (Holmes *et al.*, 1967; Jung and Kürzinger, 1968). Fig. 63 shows that both of the RNase functions are equally sensitive to irradiation in solution. The same result was obtained by Deering (1960) after UV irradiation of dry RNase. In contrast, the esterase activity in chymotrypsin (Aronson *et al.*, 1956) and trypsin (Augenstein, 1959), appears to be more radiation sensitive than the protease activity. However this question is still under discussion (see Fig. 40).

In Chapter 5.2 the target molecular weight was calculated from the D_{37} for the inactivation of numerous dry enzymes. The comparison with

116

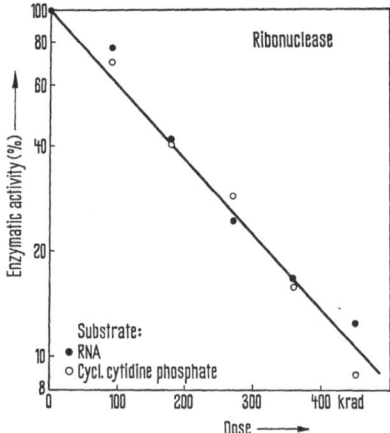

Fig. 63. Inactivation of ribonuclease in 0.5 M KCl-solution (1 mg/ml) by ^{60}Co
γ-radiation. Loss of enzymatic activity was determined using two different sub-
strates: RNA and cyclic cytidine phosphate. (Smith and Adelstein, 1965)

the actual values (Fig. 28) shows remarkable agreement. It can be concluded
from this that the radiation sensitivity of an enzyme increases with its size,
and that practically every energy loss event of 60 eV (primary event or
ion cluster) occurring at any part of a molecule leads to its inactivation.
For a long time this result caused considerable concern, since diverse changes
in enzyme molecules can be produced chemically without necessarily leading to a loss of activity.

In the determination of the radiation sensitivity of enzymes from the
slope of the dose-response curves, it is necessary to ensure that the damaged
and undamaged molecules do not have different affinities for the substrate.
Any difference would imply that the inactivation rate is a function of
substrate concentration, and the radiation sensitivity determined experimentally would, therefore, depend on the quantity of substrate used. This
problem can be resolved by a comparison of the Michaelis-Menten constants
of irradiated and unirradiated samples. According to the Michaelis theory
of the kinetics of enzyme action (see textbooks on biochemistry), the reaction
velocity v is given by:

$$v = \frac{v_{max}\,[S]}{K_m + [S]}, \tag{9.1}$$

where $[S]$ is the substrate concentration, K_m the Michaelis-Menten constant,
and v_{max} the maximum reaction velocity that can be obtained. Equation
(9.1), following the suggestion of Lineweaver and Burk (1934), can be
written in the form:

$$\frac{1}{v} = \frac{K_m}{v_{max}} \cdot \frac{1}{[S]} + \frac{1}{v_{max}}. \tag{9.2}$$

117

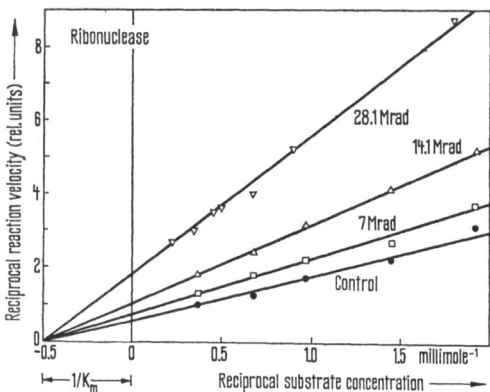

Fig. 64. Lineweaver-Burk diagram for the determination of the Michaelis-Menten constant (K_m) and the maximum reaction velocity of ribonuclease after γ-irradiation in oxygen atmosphere. (Hunt and Williams, 1964)

A straight line is, therefore, obtained if $1/v$ is plotted against $1/[S]$ ("Lineweaver-Burk diagram"). Extrapolation of this straight line to $v = \infty$, (i. e. $1/v = 0$), gives

$$1/[S]_{v=\infty} = -1/K_m. \tag{9.3}$$

The intercepts at the abscissa and the ordinate give the reciprocal of the Michaelis-Menten constant, and $1/v_{max}$ respectively. Fig. 64 shows the application of this method to dry RNase exposed to various doses of γ-radiation in the presence of oxygen; v_{max} decreases with increasing degrees of inactivation since the amount of active enzyme decreases, while the value of K_m is independent of dose. This shows that irradiated RNase has the same substrate affinity as the unirradiated enzyme. The same result has been obtained with RNase irradiated in vacuo (Hunt and Williams, 1964), and in aqueous solution (Smith and Adelstein, 1965).

In contrast, it is found that the D_{37} of DNase (Okada and Fletcher, 1962) or chymotrypsin (Mee, 1964) depends on the substrate concentration after irradiation in solution, while the action of atomic hydrogen on chymotrypsin solutions does not alter the substrate affinity (Mee et al., 1964).

9.3. Radiation-Induced Radicals

The molecular changes in irradiated enzymes will now be considered. It is advisable to begin by discussing the radicals produced by irradiation of dry enzymes. Like many other organic radicals, these are stable for long periods of time at room temperature and can easily be observed using electron spin resonance (ESR).

First of all, the ESR-spectra of enzymes with sulphur-containing amino acids, such as cystine, cysteine and methionine, will be considered qualita-

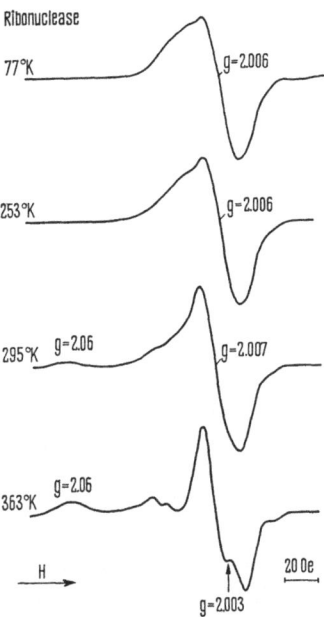

Fig. 65. ESR-spectra of ribonuclease after irradiation at 77 °K and measurement at the specified temperatures. (Copeland *et al.*, 1968)

tively; RNase, trypsin, lysozyme and pepsin belong to this group. The following interesting observation is made (Fig. 65). When RNase is irradiated at 77 °K, a wide asymmetrical resonance line is recorded at this temperature (for technical reasons, the first derivative of the resonance line is recorded, as shown in this figure). The contribution of different types of radicals to the ESR-signal at low temperatures leads to this non-specific spectrum, as these radicals are produced by the random absorption in different parts of the molecule. When the same sample is warmed to room temperature, then the form of the signal changes, and after some minutes or hours it assumes a quite specific shape caused by the superposition of at least two components. One of these components is a doublet, from which it can be assumed that it arises from an α-hydrogen on a polypeptide chain. Since polyglycine and simple glycine-containing peptides give the same spectrum, it is generally attributed to the following radical:

$$\cdots -\underset{\underset{\text{H}}{|}}{\overset{\overset{\text{H}}{|}}{\text{N}}} - \overset{\cdot}{\text{C}} - \overset{\overset{\text{O}}{\|}}{\text{C}} - \cdots \qquad (9.4)$$

However, this supposition must be qualified since certain other polyamino acids also give doublets (Drew and Gordy, 1963). The second component is

a small broad peak extending to low magnetic field which is probably due to
a sulphur radical:

$$
\cdots -\overset{\displaystyle \overset{H}{|}}{N} - \overset{\displaystyle \overset{H}{|}}{\underset{\displaystyle \underset{\underset{\displaystyle \dot{S}}{|}}{\underset{\displaystyle H-C-H}{|}}}{C}} - \overset{\displaystyle \overset{O}{\|}}{C} - \cdots \tag{9.5}
$$

This shows that the location of the primary radicals produced by the
absorption of radiation changes until finally most of the spin is located on
sulphur atoms or glycinic groups. In principle, two explanations can be
given for this phenomenon of intramolecular energy transfer, or rearrangement. On the one hand, the migration of radiation-induced "electron-holes"
along the polypeptide chain could be considered as an effective mechanism,
or alternatively it could be due to the "migration" of a radical site produced by the removal of a hydrogen atom or an amino acid side chain.
It should also be remembered that, in the discussion of intermolecular spin
transfer (Chapter 6.5), a radical localized at a sulphur atom was identified
(Fig. 46). Sulphur therefore appears to be a particularly suitable site, energetically, for the stabilization of a radical.

The G-values for the *yield of radicals* in dry irradiated enzymes lie
between 1 and 7 (Müller, 1962); i. e. 15 to 100 eV are required for the
formation of an observable radical. Since this energy is of the same order
of magnitude as the energy required for inactivation, and as furthermore
the temperature-dependence of both phenomena is often similar (Fig. 50),
it is quite probable that a large proportion of the radiation induced inactivation is a result of the formation of primary radicals.

In view of the fact that the nature of the glycine-like radicals is not
yet completely understood, the experiment of Riesz and colleagues (1966)
is of special interest. These authors exposed irradiated RNase to an atmosphere of tritiated H_2S. The radiation-induced radicals reacted according
to the following equation:

$$ M^{\cdot} + {}^3H_2S \to M^3H + {}^3HS^{\cdot}. $$

The RNase was then hydrolyzed and the individual amino acids were
separated. In contrast to what had been expected, most of the radioactivity
was not found in glycine, but (in decreasing amounts) in lysine, methionine,
proline and histidine, and in low concentrations in phenylalanine, isoleucine
and valine. According to these results, after intramolecular spin migration
the radicals will not be located on glycine in particular, but rather on a
range of several amino acids; these amino acids being, in general, those
that are destroyed with particularly high probability during irradiation
(see Table 10).

9.4. Changes in Irradiated Enzyme Molecules

The next step in this investigation of the action of radiation on enzymes is to identify the structural and molecular changes. It has been shown in Chapter 9.1 that the amino acid sequence (i. e. the primary structure of the enzyme) determines its conformation, and therefore its enzymatic activity, in an essentially unambiguous manner. It is, therefore, logical to enquire which of the amino acids are altered in irradiated ribonuclease, since an alteration in the primary structure results in a change of the conformation, which in most cases leads to the loss of enzymatic activity.

Changes in the primary structure can be determined by amino acid analysis. In the upper half of Table 10, the amino acids changed in RNase irradiated under a variety of conditions are listed. According to Augenstein (1958) the cystine content of the irradiated enzyme decreases. Since then numerous publications have expressed the view that the breakage of disulphide bonds may lead to the loss of enzymatic activity. However, this mechanism is of little significance, as will be shown later. It is interesting to note that the same amino acids are changed by the exposure of RNase solutions to atomic hydrogen and to γ-radiation. From the results discussed here, the overall conclusion was drawn that the loss of enzymatic activity of RNase is due to the change of some specific amino acids. This point will be considered again in Section 9.7.

A certain amount of information about *changes in the secondary structure* of irradiated enzymes is also available. Irradiation certainly leads to the unfolding of the molecules. This is shown by changes in the optical absorption spectrum, optical rotation, sedimentation coefficient, viscosity, and the number of hydrogen atoms available for exchange with deuterium,

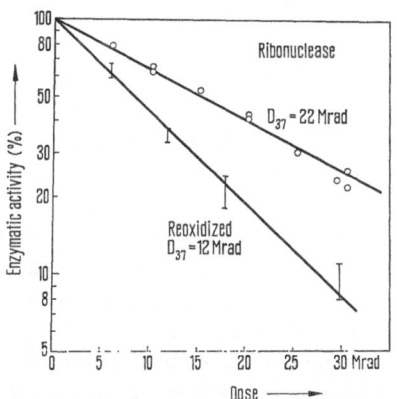

Fig. 66. Inactivation of dry ribonuclease by ^{60}Co γ-radiation in oxygen atmosphere. The activity was tested either immediately after irradiation, or after reduction followed by reoxidation of the irradiated RNase. (Haskill and Hunt, 1965)

121

as well as by the ease with which the irradiated enzymes can be digested by other enzymes. For some years, there were indications that in spite of the observed exponential dose-response curves, enzyme inactivation is not purely an "all-or-nothing" effect. Haskill and Hunt (1965) showed that *latent lesions*, not directly related to the active center, are present in active enzyme molecules after irradiation. As is shown in Fig. 66, the D_{37} for the inactivation of RNase by γ-radiation in the presence of oxygen was 22 Mrad, while for samples which had been reduced and reoxidized after irradiation it was only 12 Mrad. Thus almost half of the enzymatic activity of the irradiated molecules was lost in the post-irradiation treatment. This indicates that there are latent lesions in a fraction of those RNase molecules that are still active after irradiation, and these lesions prevent the correct refolding of a reduced molecule.

9.5. Separation and Identification of Irradiation Products

Although the experimental observations described above have given indications of many of the physico-chemical and chemical changes induced in enzymes by ionizing radiations, they are not quite adequate to provide a complete picture of the processes connected with inactivation. In order to obtain this, the irradiation products must be investigated after being separated from the unchanged molecules. This procedure may be successfully carried out chromatographically using the dextran-gel "Sephadex". The separating action of Sephadex is a result of the diffusion of small molecules into the gel, which leads to their elution from the column being slower than that of the large molecules.

Fig. 67 shows an elution pattern for RNase. The unirradiated enzyme (control) has a main peak (I_n) of native monomeric RNase, and a smaller peak (II_n) representing dimers produced during the purification of the enzyme. If RNase is irradiated in an aerated aqueous solution, two new peaks $(I_d$ and $II_d)$ are observed in the elution pattern. With increasing dose, the peak I_n decreases while the peak II_d increases steadily. The component I_d initially increases with dose, and is converted to the component II_d at doses above 1 Mrad. This component, I_d, is not observed during the irradiation of solutions under nitrogen, when with increasing degrees of inactivation, the active enzyme I_n is converted almost exclusively to the component II_d. This component still retains a small amount of enzyme activity, which is 3 times greater after irradiation under nitrogen than after irradiation in the presence of air (Jung and Schüssler, 1966).

It is possible to estimate the *molecular weight* of the various components from their relative positions in the elution pattern (Whitaker, 1963). The following values are obtained for the maxima in Fig. 67: I_n 14,000, I_d 18,000—20,000, II_n 28,000 and II_d 30,000—35,000. Comparative investigations using an analytical ultracentrifuge show, however, that I_n and I_d

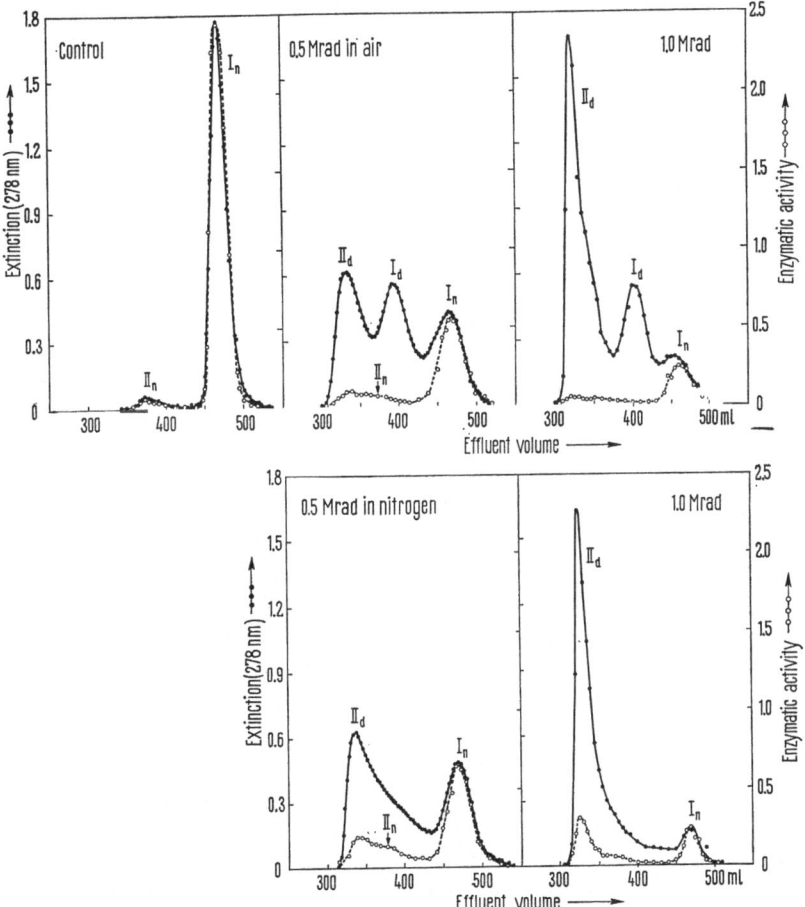

Fig. 67. Chromatographic separation of 90 mg of ribonuclease after irradiation in aqueous solution (5 mg/ml) with ^{60}Co γ-radiation. — ● — Extinction at 278 nm. — ○ — Enzymatic activity; one unit corresponds to an activity of 1 mg RNase/ml. (Schüssler and Jung, 1967)

are monomers ($MW = 14,000$), whereas II_n and II_d are dimers with a molecular weight of approximately 28,000 (Schüssler and Jung, 1967): I_d and II_d are therefore eluted from the column earlier than would be expected from their molecular weights. This indicates that these components consist of unfolded molecules which occupy a larger volume than the native molecules of the same molecular weight. These observations explain the nomenclature used: the Roman numbers give the number of molecules involved in the aggregate, while the indices n and d stand for native and denatured respectively.

The elution patterns obtained after *irradiation of dried RNase* are very similar to those shown in Fig. 67. In this case, also, the denatured monomer I_d is obtained in the presence of oxygen while after exposure in *vacuo* this component is very weak. During anaerobic irradiation, most of the native RNase is converted to denatured aggregates (Haskill and Hunt, 1967 b; Jung and Schüssler, 1968). These consist mainly of dimers, but also contain small quantities of trimers that can be separated from dimers by chromatography using Sephadex at pH 2.1 (Haskill and Hunt, 1967 b). In addition, higher-order aggregates are formed during the irradiation of dry RNase, which is reflected by the fact that a fraction of the irradiated material becomes insoluble (Jung and Schüssler, 1968).

Although the physical properties such as optical absorption, sedimentation rate, electrophoretic properties, etc. of the component I_n in Fig. 67 do not change significantly (Haskill and Hunt, 1967 a), and although it also remains enzymatically active, it nevertheless contains some damaged amino acids (Table 9), and its properties are altered following reduction and re-oxidation (Haskill and Hunt, 1967 a). In contrast, numerous physico-chemical differences are found between the *denatured components* and native RNase. As an example of this, Fig. 68 b shows the difference between the optical densities of the component I_n from the unirradiated control, and of the component II_d obtained by anaerobic irradiation of a dilute solution. This "difference spectrum" shows minima at 235, 279 and 286 nm. A very similar curve is obtained following the acid-denaturation of RNase at pH 1.4 (see Fig. 68 a), after irradiation of dry RNase (Ray and Hutchinson, 1967), and from solutions of RNase that have not been separated chromatographically following irradiation (Smith and Adelstein, 1965).

The reason why anaerobic irradiation produces only small quantities of denatured monomers (I_d; see Fig. 67) must now be considered. It can be explained by assuming that radicals are involved in the dimerization process, and that these radicals form peroxy-radicals in the presence of oxygen:

$$M^{\cdot} + O_2 \rightarrow MO_2^{\cdot}, \tag{9.7}$$

thereby inhibiting the dimerization through the interaction of two radicals:

$$M^{\cdot} + M^{\cdot} \rightarrow M - M . \tag{9.8}$$

The reaction (9.7) cannot take place in a nitrogen atmosphere; the RNase, according to (9.8) is converted almost exclusively to the denatured dimer (II_d). These considerations probably also apply in the dry state, as the irradiation of dry ribonuclease produces elution patterns similar to those found after the irradiation of solutions. However, the question still remains whether the dimerization occurs in the dry state, or only when the material is dissolved. There are some observations that support the second possibility: those amino acids containing a radical group after irradiation, which

124

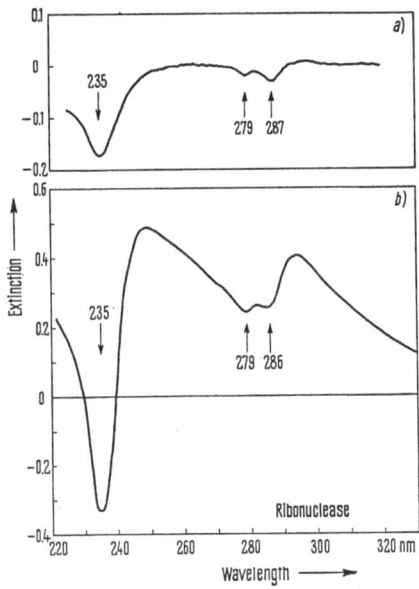

Fig. 68. a. Difference spectrum between an RNase sample denatured at pH 1.4 and native RNase at pH 5.4, measured at a concentration of 0.086% and a light path of 1 cm. (Glazer and Smith, 1961). b. Difference spectrum between component II_d (0.5 Mrad γ-radiation in anaerobic solution; concentration 5 mg/ml) and component I_n from an unirradiated control, measured at 1 mg/ml and 1 cm light path. (Jung and Schüssler, 1967)

can react with ^3H_2S (experiments by Riesz *et al.*, 1966; see Chapter 9.3), also show the greatest decrease on amino acid analysis. Furthermore, after irradiation of dry RNase in a H_2S atmosphere, no radicals are found using ESR-spectroscopy (Hunt and Williams, 1964), since they react with H_2S before the end of the irradiation; in agreement with the assumption made above no aggregation occurs in a H_2S atmosphere (Haskill and Hunt, 1967 b).

9.6. Amino Acid Analysis

Separation of the various irradiation products of RNase by Sephadex chromatography allows amino acid analysis of the individual components, thus revealing the changes in the primary structure which distinguish active from inactive molecules. The analysis of component I_n (see Fig. 67), which on considerations of the hit theory would be expected to contain only unchanged molecules, is of particular interest in this context. Table 9 shows the amino acid composition of the native monomer I_n after irradiation of ribonuclease solutions under nitrogen and under air. It is surprising to find that there are pronounced changes in the amino acid composition. After irradiation in the presence of nitrogen, the content of the amino acids cystine, methionine, tyrosine, phenylalanine, lysine and histidine is reduced, while there is a small increase in glycine content. This indicates that some of the damaged amino acids are converted to glycine by the loss of their side chains. During irradiation in aerobic conditions, the same amino acids are destroyed with one exception, cystine, which shows the greatest decrease in anaerobic conditions, but now exhibits only a small change. These results show that after irradiation in aqueous solutions, the molecules of the component I_n may be changed extensively without losing their enzymatic

Table 9. *Amino acid composition of component I_n after irradiation in aqueous solution (5 mg/ml) under nitrogen and under air. This component consists of enzymatically active ribonuclease. (Schüssler and Jung, 1967)*

Amino acid	Theoret. value	Control	Irradiated in nitrogen		in air	
			0.5 Mrad	1.0 Mrad	0.5 Mrad	1.0 Mrad
Asparagine	15	14.81	14.87	14.74	15.08	14.85
Threonine	10	10.13	10.07	9.82	10.18	10.03
Serine	15	14.99	14.68	14.81	15.04	15.15
Glutamic acid	12	12.05	12.12	12.02	12.09	12.15
Glycine	3	2.98	3.21	3.73	3.37	3.50
Alanine	12	12.01	12.00	12.02	11.79	11.64
Valine	9	8.82	8.83	8.82	8.81	8.89
1/2 Cystine	8	7.98	6.44	5.71	7.71	7.66
Methionine	4	3.77	3.63	3.43	3.55	3.32
Isoleucine	3	2.30	2.24	2.32	2.42	2.23
Leucine	2	2.03	2.06	2.07	2.08	2.07
Tyrosine	6	6.12	5.40	5.41	5.26	4.88
Phenylalanine	3	2.92	2.89	2.73	2.84	2.70
Lysine	10	9.95	9.71	9.35	9.77	8.92
Histidine	4	3.85	3.66	3.56	3.41	3.33
Arginine	4	3.99	3.95	3.83	3.87	3.71

Table 10. *Amino acids damaged in ribonuclease after irradiation under various experimental conditions*

Irradiation	Conditions	Component	Altered Amino Acids					Authors
Solution	—	—	cys			phe		Augenstein, 1958
Solution	Helium	—	cys	met	tyr			Hayden and Friedberg, 1964
Solution	Air	—			tyr			Smith and Adelstein, 1965
Solution	Air	—	cys	met	tyr	phe	lys	Slobodian and Fleisher, 1966
Solution/H_2	Air	—	cys	met	tyr	phe	his	Holmes, Navon, and Stein, 1967
Dry	Vacuum	—	cys	met				Augenstein and Grist, 1962
Dry	Vacuum	—	cys					Hunt and Williams, 1964
Radicals+H_2S	Vacuum	—		met		phe[b]		Riesz, White, and Kon, 1966[c]
Solution	N_2	I_n	cys[a]	met	tyr	phe	lys	Schüssler and Jung, 1967[d]
Solution	N_2	II_d	cys[a]	met	tyr	phe	lys	Schüssler and Jung, 1967[d]
Solution	Air	I_n	cys[b]	met	tyr	phe	lys	Schüssler and Jung, 1967[d]
Solution	Air	I_d	cys[b]	met	tyr	phe	lys	Schüssler and Jung, 1967[d]
Solution	Air	II_d	cys[b]	met	tyr	phe	lys	Schüssler and Jung, 1967[d]
Dry	Vacuum	I_n	cys	met[b]	tyr	phe	lys	Jung and Schüssler, 1968[d]
Dry	Vacuum	II_d	cys	met[b]	tyr	phe	lys	Jung and Schüssler, 1968[d]
Dry	O_2	I_n	cys	met[b]	tyr	phe	lys	Jung and Schüssler, 1968[d]
Dry	O_2	I_d	cys	met[b]	tyr	phe	lys	Jung and Schüssler, 1968[d]
Dry	O_2	II_d	cys	met[b]	tyr	phe	lys	Jung and Schüssler, 1968[d]
Dry	77° K	I_n	cys	met[b]	tyr	phe	lys	Jung and Schüssler, 1968[d]
Dry	77° K	II_d	cys	met[b]	tyr	phe	lys	Jung and Schüssler, 1968[d]

[a] Large decrease. [b] Small decrease. [c] Decrease in three further amino acids: proline, isoleucine and valine. [d] Increase in glycine.

activity. The idea, derived from the exponential form of the dose-response curves, that the inactivation of enzymes is an "all-or-nothing" process, is therefore incorrect.

Exactly the same amino acids are changed by radiation in denatured monomers as in denatured dimers (Schüssler and Jung, 1967). The selective decomposition of particular amino acids in solution could be attributed to their different reaction constants with water radicals. This correlation is particularly apparent in the case of cystine, which has an unusually high affinity for hydrated electrons (e_{aq}^-), and is decomposed more rapidly than the other amino acids when irradiated under nitrogen. The reaction is not possible in the presence of oxygen, since the e_{aq}^- is scavenged, leading to the formation of O_2^-; in agreement with this, very little cystine is changed in the presence of air. The other affected amino acids usually have higher affinities for OH radicals than those that remain unaltered (see Anbar and Neta, 1965). This correlation, however, is not as clear as with cystine. The results shown in Table 10 further restrict the validity of the hypothesis. When the different components of dry ribonuclease are irradiated, the same six amino acids are destroyed as in solution (Jung and Schüssler, 1968). It is difficult to explain this selectivity in terms of different reaction probabilities of the various amino acids. It has to be assumed that material and energy transfer processes contribute to the occurrence of the final damage (see Chapter 9.7).

Table 10 enables the observed changes in the amino acids of the different components to be compared with those obtained in a mixture of active and inactive products. The agreement is fairly good, considering the extent to which the classification of an amino acid as being intact or changed depends on the experimental accuracy and the dose.

The total number of amino acids altered can be determined as a function of dose, from the publications cited in the second part of Table 10. The sum of the experimentally determined values for cystine, methionine, tyrosine, phenylalanine, lysine and histidine in the unirradiated control is compared with the sum of these amino acids in the various separated components. The difference gives the number of damaged amino acids. Glycine is not included, because its increase is caused by the destruction of other amino acids. The reduction of cystine content in anaerobically irradiated solutions of RNase cannot be included either, since this effect is caused by selective attack of e_{aq}^-. Fig. 69 shows the total number of changed amino acids in the various components as a function of the relative dose D/D_{37} (i. e. in relation to the same degree of enzyme inactivation). This gives the interesting result that the values obtained for the component I_n after irradiation of RNase in solution, in vacuo, under oxygen, and at 77 °K, all lie on a common straight line (curve n). This means that the decomposition of amino acids parallels the loss of enzymatic activity under these

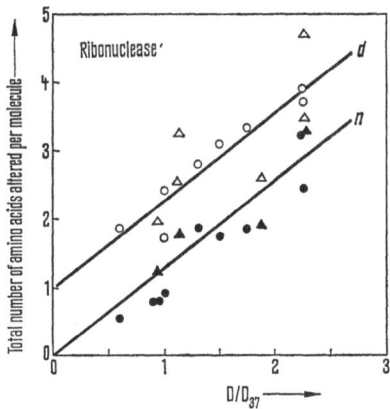

Fig. 69. Sum of amino acids damaged per RNase molecule in the different components. n = enzymatically active ribonuclease; d = inactive radiation products. ● Component I_n, ○ Component I_d and II_d after irradiation in the dry state, under oxygen, in vacuo, and at 77 °K. ▲ Component I_n, △ Component I_d and II_d after irradiation in solution (5 mg/ml) in air and under nitrogen. (Schüssler and Jung, 1967; Jung and Schüssler, 1968)

experimental conditions, although the corresponding 37%-doses differ by as much as a factor of 5,000. The values for the amino acid changes obtained for the denatured components I_d and II_d also lie on a common straight line, having the same slope as curve n, and intercepting the ordinate at 1. This curve is, of course, not defined mathematically for $D=0$, since with no irradiation there is no denatured component.

In the enzymatically active component, 1.25 amino acids are destroyed per inactivated molecule (i. e. when $D/D_{37}=1$), whereas the denatured component already contains 2.25 damaged amino acids. This implies that if one of the six amino acids mentioned is destroyed, then there is a probability of 0.45 that the molecule will be inactivated, whereas in 55% of all cases the enzyme retains its activity (Jung and Schüssler, 1968).

9.7. Mechanisms of Inactivation

The results presented in the previous section will now be utilized to give an understanding of the effects induced in enzymes by ionizing radiation. In order to explain the specificity of amino acid decomposition shown in Table 10, as well as the ESR-results of Fig. 65, it is necessary to assume that the primary lesions produced by the absorption of radiation energy, although initially distributed at random throughout the whole molecule, may finally become stabilized at favourable sites by *intramolecular energy transfer*, or rearrangement. These sites will predominantly be divalent sulphur or conjugated ring systems (Holmes *et al.*, 1967). This mechanism is essentially independent of the type of primary lesion, which explains

129

why it is always the same amino acids that are altered in the irradiation of solutions, by the action of hydrogen atoms in solution, and during irradiation in the dry state under a variety of experimental conditions.

The fact that fewer amino acids are altered per unit dose at 77 °K than at room temperature is due to the reduction in the indirect action of hydrogen atoms at low temperatures (Chapter 7.2). The observation that only the number, and not the type, of detectable amino acid alterations is affected by the presence of oxygen, can be explained in a similar manner. According to the hypothesis discussed in Chapter 8.1, the presence of oxygen largely inhibits the restitution reactions (8.2) and (8.3), and therefore increases the overall number of damaged molecules.

The *dimerization* produced by the interaction of two radicals has already been described in section 9.5 (equation 9.8). Some of this aggregation is produced by disulphide exchange, i. e. the sulphur radical on a cystine residue reacts with a similar lesion in another molecule, thereby linking two RNase molecules by a disulphide bridge. This is reflected in the fact that a part of the aggregates can be converted into monomers by the reduction of the disulphide bridges. In the presence of oxygen, 10 to 60 per cent of the aggregates are derived from disulphide exchanges, and possibly even more in vacuo (Haskill and Hunt, 1967 c).

The dimerization results from alterations in amino acids, but it is not the cause of the loss of enzymatic activity. This is shown by irradiation under H_2S, when the formation of the dimers is completely inhibited, with no resultant protective effect; the radiation sensitivity of RNase is actually 30% higher than in vacuum (Hunt and Williams, 1964). One of the two participants in a dimerization process may occasionally retain its enzymatic activity, which would explain the enzyme activity observed in denatured dimers (see Fig. 67).

What process causes the unfolding of the molecule that occurs in about 50% of the cases in which an amino acid has been altered? Some of the inactivated molecules have *breaks in their peptide chains*. This cannot be recognized from the elution pattern (see Fig. 67), as in most cases no fragments are formed initially, the two parts of the molecule being held together by the disulphide bridges (see Fig. 62). Such a "masked" break in the polypeptide chain can, however, be detected by reducing the disulphide bridges of the irradiated component before the chromatographic separation. Breaks are found in denatured components, but not in the enzymatically active molecules of component I_n (Haskill and Hunt, 1967 c; Ray and Hutchinson, 1967). This means that a break in the peptide chain, together with an alteration of an amino acid, will lead in every case to the loss of enzymatic activity. Breaks alone do not necessarily cause inactivation, since several breaks can be induced in a RNase molecule by enzymatic attack, without loss of the enzymatic activity (G. Pfleiderer, personal communication). The breaks do not occur in a few characteristic sites, but are distributed throughout the molecule. The different fragments can be separated by starch-gel electrophoresis, which gives at least eight poorly resolved peaks (Haskill and Hunt, 1967c). A break in a polypeptide chain should, according to Garrison and Weeks (1962), be coupled with the formation of one carbonyl bond and one amide group; however, no quantitative agreement has yet been obtained on this point.

However, not all the inactivated molecules contain a break in the polypeptide chain. The fraction varies between 5% (Ray and Hutchinson, 1967) and 50% (Haskill and Hunt, 1967c), depending on the experimental conditions. There must, therefore, be some other mechanism responsible for most of the transitions from the active to the inactive state. Unfolding of the RNase molecule probably occurs when the altered amino acid is a part of a *hydrophobic bond*. If the hydrophobic character is destroyed, or alternatively if the newly formed residue carries a different charge, then the polypeptide chain will, in many cases, assume a different configuration. The extent to which the alteration of a single amino acid may affect the configuration of a protein is shown by the example of hemoglobin S. The properties of this mutant, observed in sickle-cell anaemia, differ significantly from those of normal haemoglobin; the only difference in its primary structure, however, consists of the replacement of a hydrophobic valine by a charged amino acid residue, glutamic acid.

Under these conditions, a continuous transition from active to inactive molecules should occur, which has already become apparent from Fig. 69. The magnitude of the damage will depend on the site of the altered amino acid within the molecule, the charge on the newly formed residue, and the hydrophobic properties of this residue. If, after a dose of 12 Mrad in oxygen, the various components of RNase are reduced and then reoxidized, about one quarter of the molecules of the type I_n lose their enzymatic activity in this procedure, while the activity of components I_d and II_d is reduced by 60—70% (Haskill and Hunt, 1967a). This means that there are changes in the primary structure of the irradiated molecules that are not sufficiently severe to cause unfolding (and therefore inactivation), but which are adequate to prevent the correct refolding of the molecule after denaturation. An example of the sensitivity of the RNase molecule to changes in its hydrophobic region is given by the experiments of White (1964). According to these investigations, the enzyme activity is not altered by the insertion of two strongly aromatic side chains; the modified molecule is, however, completely unable to regain its original conformation after reduction.

During the irradiation of dry ribonuclease in vacuo, a small portion of the molecules is inactivated by a mechanism proposed by Platzman and Franck (1958). The introduction of an electrostatic charge into the molecule by an ionization gives rise to an inhomogeneous electrostatic field which polarizes certain polar side groups, and in this process a number of hydrogen bonds are broken. This produces a change in the conformation, which leads to inactivation. The change can be reversed by unfolding and then reconstituting the molecule. This is shown by the results of Haskill and Hunt (1967 a), according to which the enzymatic activity of the monomer component I_n of an anaerobically irradiated RNase sample increases by 20% after reduction and reoxidization, while in the denatured irradiation products, the enzymatic activity is strongly reduced by these procedures. The experiments show that only a small part of the molecules irradiated in vacuo are damaged by this mechanism, while in the presence of oxygen the ionized molecules react and are thereby changed irreversibly.

The inactivation scheme developed in this section describes a large proportion of the physico-chemical and chemical changes occurring after irradiation of RNase, as well as the processes involved in the inactivation of this enzyme. This scheme is essentially consistent: it agrees with the result of electron spin resonance spectroscopy, it describes the sensitizing action of oxygen and the protection by low temperatures, as well as the selective decomposition of several amino acids under a variety of experimental conditions. Furthermore, it explains why the absorption of radiation in any part of the molecule has a high probability of leading to inactivation, while chemically induced damage has no effect in many cases. The attack of chemical agents is mostly concentrated at the surface, and therefore does not affect the bonds within the molecule and the molecular conformation. Radiation energy, in contrast, regardless of the part of the molecule to which it is initially transferred, may reach the interior by processes of energy migration, thereby causing inactivation. Some parts of this picture, drawn for the specific case of ribonuclease, may differ for other enzymes. It may, however, be assumed that the overall picture is correct in many aspects. The extent to which this assumption is true can only be decided when other enzymes have been investigated in as much detail as ribonuclease.

References

Anba, M., Neta, P.: Int. J. appl. Radiat. Isotopes **16**, 227 (1965).
Aronson, D., Mee, L., Smith, C. L.: In: Progress in radiobiology. Eds.: J. S. Mitchell, B. E. Holmes, and C. L. Smith. Edinburgh: Oliver & Boyd 1956, p. 61.
Augenstein, L. G.: In: Symposium on information theory in biology. Eds.: H. P. Yockey, R. L. Platzman, and H. Quastler. New York: Pergamon Press 1958, p. 287.
— Science **129**, 718 (1959).
— Grist, K.: Cited by L. G. Augenstein. Adv. Enzymol. **24**, 359 (1962).
Copeland, E. S., Sanner, T., Pihl, A.: Radiat. Res. **35**, 437 (1968).

Deering, R. A.: Arch. Math. Nat. (Oslo) **55**, Nr. 5 (1960).

Drew, R. C., Gordy, W.: Radiat. Res. **18**, 552 (1963).

Garrison, W. M., Weeks, B. M.: Radiat. Res. **17**, 341 (1962).

Glazer, A. N., Smith, E. L.: J. biol. Chem. **236**, 2942 (1961).

Haskill, J. S., Hunt, J. W.: Biochim. Biophys. Acta **105**, 333 (1965).

— — Radiat. Res. **31**, 327 (1967a).

— — Radiat. Res. **32**, 606 (1967b).

— — Radiat. Res. **32**, 827 (1967c).

Hayden, G. A., Friedberg, F.: Radiat. Res. **22**, 130 (1964).

Holmes, B. E., Navon, G., Stein, G.: Nature **213**, 1087 (1967).

Hunt, J. W., Williams, J. F.: Radiat. Res. **23**, 26 (1964).

Jung, H., Kürzinger, K.: Radiat. Res. **36**, 369 (1968).

— Schüssler, H.: Z. Naturforsch. **21 b**, 224 (1966).

— — unpublished results (1967).

— — Z. Naturforsch. **23 b**, 934 (1968).

Kartha, G., Bello, J., Harker, D.: Nature **213**, 862 (1967).

Lineweaver, H., Burk, D.: J. Amer. chem. Soc. **56**, 658 (1934).

Mee, L. K.: Radiat. Res. **21**, 501 (1964).

— Navon, G., Stein, G.: Nature **204**, 1056 (1964).

Müller, A.: In: Biological effects of ionizing radiation at the molecular level. Vienna: Internat. Atomic Energy Agency 1962, p. 61.

Okada, S., Fletcher, G.: Radiat. Res. **16**, 646 (1962).

Platzman, R. L., Franck, J.: In: Symposium on information theory in biology. Eds.: H. P. Yockey, R. L. Platzman, and H. Quastler. New York: Pergamon Press 1958, p. 262.

Ray, D. K., Hutchinson, F.: Biochim. Biophys. Acta **147**, 357 (1967).

Riesz, P., White, F. H., Kon, H.: J. Amer. chem. Soc. **88**, 872 (1966).

Schüssler, H., Jung, H.: Z. Naturforsch. **22 b**, 614 (1967).

Slobodian, E., Fleisher, M.: Biochemistry **5**, 2192 (1966).

Smith, T. W., Adelstein, S. J.: Radiat. Res. **24**, 119 (1965).

Smyth, D. G., Stein, W. H., Moore, S.: J. biol Chem. **238**, 227 (1963).

Whitaker, J. R.: Analyt. Chem. **35**, 1950 (1963).

White, F. H.: J. biol. Chem. **239**, 1032 (1964).

Chapter 10. Physico-Chemical Changes in Irradiated Nucleic Acids

The nucleic acids have a fundamental role in the maintenance of vital processes. While deoxyribonucleic acid (DNA) carries genetic information, the various ribonucleic acids fulfil important functions in the realization of this information (see Chapter 11.1). The key position in biological processes occupied by nucleic acids made the investigation of the action of radiation on DNA and RNA the central theme of molecular radiation biology. As in the case of enzymes, the main problem is to correlate loss of biological function with the occurrence of physical and chemical changes, and thereby to gain an understanding of the inactivation mechanism. There are numerous biological functions of the nucleic acids that are accessible to measurement, as well as physico-chemical changes induced by irradiation, so that the simultaneous discussion of both of these facets would affect the clarity of this presentation. The physico-chemical and chemical changes occurring in irradiated nucleic acids will, therefore, be considered first, and in the three succeeding chapters an attempt will be made to correlate these changes with the inhibition of certain biological functions. This will not always be simple, as in most cases only one of these effects has been examined by a particular author. The attempts to relate functional with physico-chemical changes have only developed during recent years, especially in the work with bacteriophages. Since ribonucleic acid has been examined far less extensively than deoxyribonucleic acid, the considerations of this chapter will be restricted exclusively to DNA.

10.1. The Structure of DNA

A better understanding of the experimental methods and results is likely to be gained after reviewing some of the available information about the structure of DNA molecules (Fig. 70). In double-stranded DNA, two polynucleotide chains are wound around each other in a double helix, the base sequence being specific for each type of DNA. Between the individual base pairs there are two or three hydrogen bonds holding the two strands together. Both strands can be separated partially or completely by heating the DNA in the presence of various chemicals; this is referred to as partial

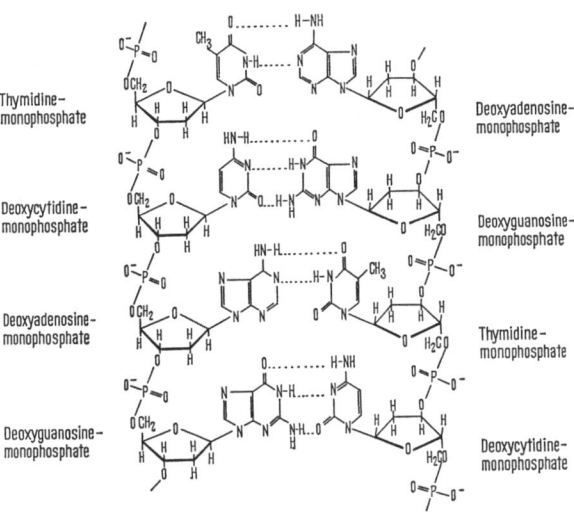

Fig. 70. Structure of double-stranded DNA with hydrogen bonds

or complete *denaturation*. This must be distinguished from *degradation,* which generally refers to a shortening of the chain length of the molecules by scission of one or both of the polynucleotide strands.

The diameter of the DNA double helix is approximately 20 Å. Since the total genetic information of bacteriophages or bacteria is condensed into a single DNA molecule, this often reaches a considerable length. The genome of the bacterium E. coli consists of a double-stranded DNA molecule in the form of a closed ring, having a molecular weight of $3 \cdot 10^9$ Dalton and a length of about 1 millimeter. This is equivalent to a piece of string 1 mm in diameter and 5 km long!

A series of possible changes that could occur during the irradiation of DNA can be visualized from Fig. 70. First of all, free radicals are formed in various parts of the DNA molecule, and these can be investigated in dry DNA with the aid of ESR-spectroscopy. The production of this primary reactive species may result in chemical changes, such as deamination or dehydroxylation, scission of the base-sugar bond, oxidation of the sugar, or release of the phosphate group. These reactions lead finally to changes in the macromolecular structure of DNA. In this context, it is necessary to distinguish between single strand breaks, double strand breaks (i. e. coincident or adjacent breaks on the two nucleotide strands) and cross-linking of two or more DNA molecules. These alterations in the macromolecular structure will also affect the hydrogen bonds; the investigation of these changes will further the understanding of the radiation effects in DNA, although the rupture of hydrogen bonds in itself does not in general cause biological damage.

10.2. Radiation-Induced Radicals

The primary phases of the chemical changes in irradiated nucleic acids are characterized by the occurrence of free radicals which can be observed using ESR-spectroscopy. Such a vast amount of ESR-literature on DNA and its components has been accumulated, that it is not possible within the scope of this book to present a comprehensive review of the field. Therefore, only some representative results will be discussed, and further information can be obtained from the review articles of Zimmer and Müller (1965) and Müller (1967).

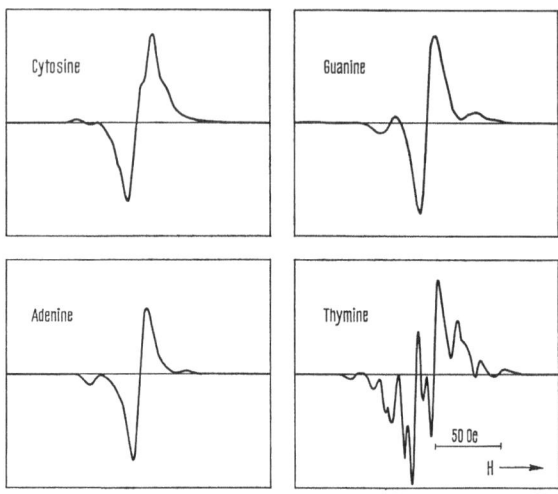

Fig. 71. ESR-spectra of DNA bases after ^{60}Co γ-irradiation. Irradiation and measurement in vacuo at room temperature. (Köhnlein, 1963)

Quantitative ESR-spectroscopy is used to determine the yields of radicals induced in DNA and its constituents. The ESR-spectra obtained at room temperature for the four most important DNA bases are shown in Fig. 71. As was mentioned in Chapter 9.3 (see Fig. 65), for purely technical reasons the first derivative of the absorption signal is recorded. The signals of the bases have little hyperfine structure, with the exception of thymine the main component of which consists of a characteristic eight-line spectrum. Fairly similar spectra are obtained by irradiation at low temperatures, followed by subsequent measurement at room temperature. The number of radicals present in the sample is derived from the intensity of the ESR-spectra (by integrating the spectra twice). The yield of radicals increases in the order base < nucleoside < nucleotide, as shown in Table 11. The reasons for this will become apparent in the discussion of qualitative ESR-spectroscopy. The radical yield and the spectra obtained from DNA depend to a large extent on the kind of DNA and the preparation proce-

Table 11. *Comparison of the G-values for the production of radiation-induced free radicals in nucleic acid constituents at 300 °K.* (Müller, 1964)

Bases, Sugar	G-value	Nucleosides	G-value	Nucleotides	G-value
Adenine	0.1	Adenosine	1.4	dAMP	2
Guanine	0.8	Guanosine	0.9	dGMP	3
Cytosine	0.4	Cytidine	1.0	dCMP	5
Thymine	0.1	Thymidine	0.4	dTMP	2
β-2-Deoxy-D-ribose	4				

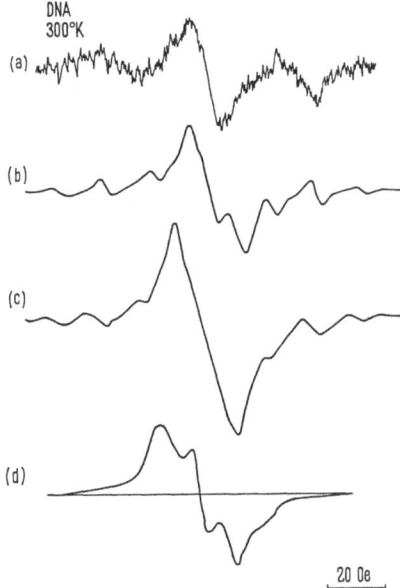

Fig. 72. ESR-spectra of various irradiated DNA preparations (Müller, 1967). a DNA from trout sperms (Dorlet *et al.*, 1962). b Calf thymus DNA (Salovey *et al.*, 1963). c Calf thymus DNA (Ehrenberg *et al.*, 1963). d DNA from T2-bacteriophage. (Müller, 1963)

dures. This is the reason why different laboratories have obtained G-values between 0.2 and 12 at 77 °K, the variation at room temperature being even larger.

Using *qualitative ESR-measurements*, attempts can be made to identify the radicals contributing to the spectra of irradiated nucleic acids. Fig. 72 shows the ESR-signals from some different DNA preparations. It is a characteristic observation that the eight-line spectrum of thymine is observed with quite different intensities, and it has been shown by several authors that this spectrum is due to an addition product of atomic hydrogen

137

to the C_6-atom (see Table 12) of thymine (e. g. Pershan *et al.*, 1964). The identification of the other components of the DNA signal shown in Fig. 72 is more difficult. The first spectrum (a) could consist, at least partly, of a triplet with a splitting of 35 Oe which could, for example, be attributed to a guanine radical (see Table 12). At least three lines are involved in the fourth spectrum (d), and the central line can be removed by certain of the preparative procedures (Müller, 1964), leaving a doublet with a splitting of about 20 Oe which can be attributed to a radical of the type $-\overset{\bullet}{\underset{|}{C}}-H$.

However, it is not clear where this radical is located on the DNA molecule. If the 8-line thymine spectrum is subtracted from the spectra (b) or (c) of Fig. 72, then a triplet with a splitting of about 20 Oe remains (Pershan *et al.*, 1964), the origin of which is also uncertain.

The difficulties in correlating the basic structure of the DNA signal with specific radicals suggests that the quantitative ESR-investigations should be extended to the *constituents of DNA*. Whenever possible, single-crystals of the compounds to be investigated are used. Such techniques allow the dependence of the ESR-spectra on orientation to be investigated. In this procedure, the crystal is rotated in the magnetic field about three mutually perpendicular axes, and the spectral splitting, for example, is measured as a function of orientation. The tensor of the hyperfine splitting is then generally specific for the radical structure. Some radicals of the DNA constituents, which have been identified using this procedure, are compiled in Table 12. A notable feature is that all these base radicals are formed by hydrogen addition, in which the 5,6-double bond on pyrimidines is affected predominantly, while addition to the purines occurs at positions 2 or 8. Deoxyribose forms a radical via the loss of a hydrogen atom and ring-opening. It is now possible to explain the above results, according to which fewer radicals are produced in bases than in nucleosides, which in turn carry fewer than nucleotides. The higher yield of radicals in nucleosides and nucleotides is due to the additional attack of the hydrogen atoms split off the sugar, i. e. to an indirect effect according to equation (6.5). A few, or no, radicals are stabilized in the sugar itself, in sharp contrast to the irradiation of free sugars.

The production of radicals by the addition of hydrogen to bases can be shown directly by exposing finely powdered samples to atomic hydrogen derived from a gas-discharge. ESR-spectra similar to those produced by the irradiation of nucleosides are obtained (Herak and Gordy, 1965). Similar ESR-signals are also obtained after the exposure of DNA to atomic hydrogen (Heller and Cole, 1965).

The question of *intramolecular spin transfer* (rearrangement) in nucleic acids, and especially in DNA, will now be discussed. That it does occur in DNA and also in its components has been shown, for example, by Müller

Table 12. *Some radicals of DNA constituents identified in single-crystals by ESR-spectroscopy. The hyperfine splitting refers to the isotropic part of the spectrum*

Crystal	Number of lines intensity ratio splitting	Radical	Authors
Deoxyadenosine-monohydrate	3 1:2:1 43.5 Oe	[structure: adenine radical with NH_2; ring atoms numbered 1 2 3 4 5 6 7 8 9]	Lichter and Gordy, 1968
Deoxyguanosine-hydrochloride	3 1:2:1 35 Oe	[structure: guanine radical with O, $H-N$, NH_2]	Dertinger, 1967 a
Cytosine-monohydrate	2 1:1 10 Oe	[structure: cytosine radical with NH_2, HO, ring atoms numbered 2 3 4 5 6, H]	Dertinger, 1967 b
	6 1:1:2:2:1:1 19 Oe	[structure: cytosine radical with NH_2, H, $O=C$, N-H]	Cook et al., 1967
Thymidine	8 1:3:5:7:7:5:3:1 20.5 Oe	[structure: thymine radical with O, $H-N$, $C-CH_3$, $O=C$, N, H]	Pruden et al., 1965
β-2-Deoxy-D-ribose	5 1:4:6:4:1 10.5 Oe	[structure: deoxyribose radical, ring atoms numbered 1' 2' 3' 4' 5', H H, HO, $O=C-OH$, H OH, H H]	Hüttermann and Müller, 1969

(1964). These materials were irradiated in the dry state at 77 °K, and the signals were measured at this temperature. After the samples had warmed up to 300 °K, the form of the spectrum changed. This shows that the DNA signal obtained at room temperature is not formed purely by a super-position of the primary spectra of the individual DNA components, but as with enzymes (see Fig. 65) rearrangement effects influence the signal.

The question of whether energy can be passed from one strand to the other through the system of hydrogen bonds within the double-stranded DNA, is of special interest in this context. Schmidt and Snipes (1967) carried out an interesting ESR-study of this problem. They irradiated a 1:1-cocrystal of 9-methyladenine and 1-methylthymine at 77 °K, which gave an non-specific ESR-signal. The spectrum obtained after warming to room temperature, was derived almost exclusively from a radical on the thymine ring. A similarly treated equimolar mixture of the two compounds gives a sum-spectrum of the adenine and thymine signals. This study illus-trates the possibility of energy or spin transfer across hydrogen bridges in a cocrystal. However, it cannot be concluded definitely from this experi-ment that energy transfer across the hydrogen bonds of the base pairs occurs within the DNA molecule. However, the observation that no sig-nificant contribution from an adenine signal can be recognized, particularly in DNA molecules showing a pronounced thymine signal, is an indication of a mechanism involving preferential energy transfer to thymine.

10.3. Chemical Changes in Irradiated DNA

While the ESR-investigations just mentioned give information about the paramagnetic nature of the processes occuring after irradiation of dry DNA, the experiments now to be described will refer to the radiolysis of DNA and its constituents in *aqueous systems*. Maintaining the same concepts, the discussion will concentrate on the action of ionizing radiation. However, a short reference to the changes induced by the action of light will be made, as they are relevant to an understanding of the results described in Chapters 12 and 13. This can only be a short summary; further information about the action of UV radiation can be found in the book of Smith and Hanawalt (1969). A detailed discussion of the changes induced in DNA by ionizing radiation will be found in the review articles by Scholes (1963), Weiss (1964), and Kanazir (1969).

a) Base Changes

UV light. Although there are only relatively few quantitative studies on the photochemistry of purine bases, there is an extensive literature on the effects of UV irradiation on pyrimidine bases. The reason for this is prob-ably the fact that the pyrimidines are destroyed by UV irradiation (wave-length 253.7 nm) approximately ten times as effectively as the purines. The corresponding quantum yields (i. e. the number of changed molecules divided by the number of absorbed UV quanta) are about 10^{-3} and 10^{-4}, respectively. It is, therefore, assumed that from the biological point of view the changes in pyrimidine bases are more important than those in the purine bases.

As well as the hydration of the 5,6-double bond of pyrimidines, which is reversible by treatment with heat or acid, and therefore probably less important biologically, the dimerization of adjacent pyrimidine bases occurs and is the most significant UV lesion. In this process, the bases are linked by carbon-carbon bonds between positions 5 and 6, creating a cyclobutane ring between the two pyrimidines (Beukers and Berends, 1960). The most frequently formed dimer is a thymine dimer, of which there are six possible isomers. So far, five of these isomers have been found in irradiated oligonucleotides (Weinblum and Johns, 1966); some of these are stable to acid hydrolysis. The dimerization cannot be reversed by heat treatment.

The biological importance of the thymine dimers is reflected in the decrease in the survival rate and transforming activity of bacteria as the number of dimers formed increases (Wacker et al., 1962; Setlow and Setlow, 1962). The rate of formation of dimers has a maximum at intermediate UV wavelengths (280 nm), since more energetic radiation (240 nm) is able to split the previously formed dimers. This is in agreement with the biological observation that the inactivation of transforming ability of DNA caused by 280 nm-radiation can be partially reversed by a second irradiation at 240 nm (see Fig. 84). Although thymine dimers are the most important and frequent lethal lesions after UV irradiation, the dimers of the other pyrimidines also contribute, as well as certain other lesions.

Ionizing Radiation. The decomposition of free bases irradiated in aqueous solutions can be detected by the decrease in optical absorption in the UV region, or using paper chromatography, as well as by specific chemical reactions (e. g. with bromine or silver salts). These experiments show that pyrimidines $(G = 1.9 - 2.1)$ are almost twice as sensitive as purines $(G = 1.1 - 1.3;$ Scholes et al., 1960). In pyrimidines the OH radicals attack predominantly at the 5,6-double bond. Under aerobic conditions, this leads to the formation of pyrimidine-hydroxy-hydroperoxide, and under anaerobic conditions by the action of two OH radicals, to the formation of a glycol. The hydroxy-hydroperoxides of uracil and cytosine are unstable, and react to form glycol and isobarbituric acid, respectively. The details and intermediate reaction steps are described by Latarjet et al. (1963). In contrast to the action of UV light, no dimers are formed by the irradiation of frozen thymine solutions with ionizing radiation (Wacker and Lochmann, 1962).

The purines have not yet been investigated in as much detail as the pyrimidines. Under anaerobic conditions, it is mainly the imidazole ring that is destroyed, while in the presence of oxygen a peroxidation of the

5,6-double bond occurs. The resulting hydroxy-hydroperoxide has not yet been isolated, possibly because it is unstable. Cyclonucleotides have also been found (Keck, 1968), and may play a part in the process of cross-linking.

If a mixture of four DNA bases is irradiated in solution, it is again found that there is twice as much decomposition of the pyrimidine bases as of the purines. A similar result is obtained with an equimolar mixture of the four nucleotides (McCargo, 1961). In summary, it can be said that the extent of base decomposition decreases in the following order: Free base > nucleoside > nucleotide > DNA, while in the previous section it was shown that the radical production increases in this order in dry DNA. Table 13 gives the G-values for the decomposition of the individual bases

Table 13. *G-values for the decomposition or release of purine and pyrimidine bases after irradiation of DNA in aqueous solution*

Base	aerobic [a]	Destroyed aerobic [b]	anaerobic [c]	Released anaerobic [d]
Adenine	0.39	0.42	0.12	0.069
Guanine	0.26	0.64	0.19	0.043
Cytosine	0.38	0.54	0.27	0.071
Thymine	0.64	0.72	0.43	0.045
Total	1.67	2.32	1.01	0.23

[a] 0.2% conc., 200 kVp X-radiation, oxygen (Scholes et al., 1960). [b] 0.2% conc., 15 MeV electrons, oxygen (Hems, 1960). [c] 0.5% conc., 15 MeV electrons, oxygen-free (Hems, 1960). [d] 0.5% conc., 15 MeV electrons, nitrogen (Hems, 1960).

in an irradiated DNA solution. It can be seen that in the presence of oxygen, twice as many bases are destroyed as under anaerobic conditions, and that thymine is more rapidly decomposed than any of the other bases. The destruction of bases occurs about 4 times more frequently than the loss of a base from irradiated DNA, under comparable experimental conditions. This last effect is not influenced by the presence of oxygen (Hems, 1960).

b) Changes in the Sugar

The release of undamaged bases is mainly the result of a chemical attack on the sugar group of the DNA molecule, and is most probably due to hydrolysis of the N-glycosidic bond (Scholes and Weiss, 1952). Under anaerobic conditions, reducing water radicals appear to attack the sugar moiety (although they have a very great affinity for the double bonds of the bases), since the G-value for the production of H_2 in irradiated nucleotide solution is $G = 1.0$ (Daniels et al., 1957), while a G-value of only 0.6 is obtained in pure water under similar conditions. According to this, a G-value of 0.4 for the abstraction of hydrogen from sugar by means of reaction (6.7) would be expected. This indirect conclusion is supported by the observation that in the irradiation of equimolar solutions of the four nucleotides, the bases are destroyed with a G-value of 1.25, and the sugars with a G-value of 0.73; i. e. 37% of the attacks are on the sugar (McCargo, 1961). This is in approximate agreement with the results of Scholes *et al.* (1960), according to which about 20% of the water radicals react with the sugar and 80% with the bases.

Damage to the sugar, according to Scholes *et al.* (1960), leads in general to a break in the nucleotide chain. This then leads to the release of 3'- or 5'-monophosphate, either directly or as a result of the damage to the sugar, forming an unstable intermediate diester which ultimately loses its damaged sugar group. A second attack by a water radical can convert this monoester group to free inorganic phosphate. The corresponding G-values are small and increase, as a result of the higher-order kinetics, with increasing dose (Scholes and Weiss, 1954). The formation of single-bond phosphate groups by the irradiation can be detected with the help of phosphomonoesterase. This enzyme, after a sufficiently long incubation, splits inorganic phosphate from all the terminal monoester groups, and this can subsequently be measured photometrically. At the same time, this procedure gives the number of radiation-induced single strand breaks; G-values of 0.4 to 0.8 are obtained for the scission of one of the polynucleotide chains by the aerobic irradiation of 0.1 and 0.5% DNA solutions, respectively (Collyns *et al.*, 1965).

In addition to these chemical changes in irradiated DNA, numerous endproducts of damaged DNA components, such as hydroperoxides, hydrogen peroxide, ammonia etc., have been found. As no statements can be made about the mechanisms by which these are formed, they will not be considered further.

10.4. Breaks in the Polynucleotide Chains

Changes in the macromolecular structure of DNA are generally detectable by changes in the molecular weight; double strand breaks and cross-links between two molecules influence the distribution of molecular weights directly. Single strand breaks, which can be detected by the enzymatic procedure mentioned above, influence the distribution of molecular weights only after denaturation of the irradiated DNA. This is achieved by heating the DNA to 90° C for 10 min, or alternatively by treating with alkali. It is then cooled rapidly to 0° C, or formaldehyde added, and neutralized in order to prevent renaturation. The formation of hydrogen bonds within the coiled-up single strands usually cannot be avoided.

A variety of different methods is commonly used for the determination of the molecular weights of macromolecules: for example, light scattering, osmosis, viscosity, sedimentation measurements, etc. Comparable results are obtained using these methods only when the test sample contains molecules with a sufficiently uniform distribution of molecular weights. This has not been the case, however, in most of the DNA samples investigated so far, since fragments with quite different molecular weights are produced during the preparation. Two averages can be derived from such distributions, the number-average molecular weight

$$M_n = \Sigma\, n_i\, M_i / \Sigma\, n_i \qquad\qquad (10.1)$$

and the weight-average molecular weight

$$M_w = \Sigma\, n_i\, M_i^2 / \Sigma\, n_i\, M_i\,, \qquad\qquad (10.2)$$

where n_i represents the number of molecules with molecular weight M_i. For a random distribution of molecular sizes (which applies approximately in many DNA preparations) the following relationship holds:

$$M_n : M_w = 1 : 2\,. \qquad\qquad (10.3)$$

Molecular weight determinations on non-uniform samples give mean values which depend on the method of measurement; results between M_n and M_w are obtained with most procedures. Osmotic measurements give the mean number M_n. However, this rather time consuming procedure is very sensitive to the presence of low molecular weight impurities, and has only rarely been applied in radiation biological experiments. The weight-average molecular weight M_w is obtained by measurements of the light scattering, while viscosity and sedimentation measurements give a result slightly below M_w. An unambiguous result is obtained when the sedimentation distributions at different concentrations, determined with the aid of an analytical ultracentrifuge, are extrapolated to zero concentration, and this "zero-concentration" distribution converted to a molecular weight distribution according to Eigner and Doty (1965). From this, the two quantities M_n and M_w, as well as the rates of breakage and cross-linking, can be calculated (cf. Hagen, 1970). As the initial sizes of the irradiated molecules may vary, the number of breaks is usually not calculated per molecule, but their frequency is given. Single strand breaks are often expressed per nucleotide (B_1), and double strand breaks per nucleotide pair (B_2). We will use this nomenclature in the following sections.

Irradiation of dry DNA in vacuo yields single and double strand breaks the frequency of which increases linearly with dose (Fig. 73). The G-values derived from these curves are $G = 0.63$ for a single break and $G = 0.11$ for a double break; i. e. single strand breaks are 5—6 times more frequent than double strand breaks. If the irradiation is carried out under oxygen, then the yield of double breaks is only marginally increased ($G = 0.16$), while the number of single strand breaks is strongly enhanced ($G = 3.4$) (Hagen and Wellstein, 1965). The fact that the presence of oxygen increases the

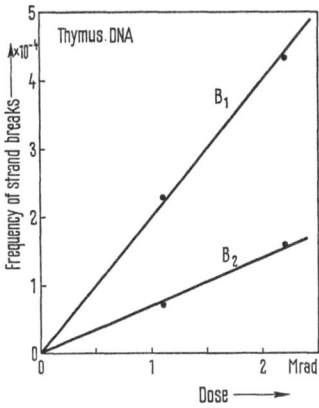

Fig. 73. Production of single and double strand breaks in dried calf thymus DNA by irradiation with X-rays in vacuo. B_1 = frequency of single strand breaks per nucleotide. B_2 = frequency of double strand breaks per nucleotide pair. (Hagen and Wellstein, 1965)

frequency of double strand breaks only marginally, has also been observed by Alexander and collaborators (Alexander *et al.*, 1961; Lett *et al.*, 1961), using calf thymus DNA, and by Freifelder (1965) using DNA from phage T7. The different responses of single and double strand breaks to the presence of oxygen has already been considered in the discussion of the oxygen effect in microorganisms (see Chapter 8.2).

The number of double strand breaks in irradiated dry DNA increases linearly with dose (Fig. 73), as has already been seen. This implies that the double break in the dry system is produced by a single energy-loss event, and not by the random coincidence of two adjacent single breaks: the probability of this process should therefore increase with the LET of the radiation. In confirmation of this, Dewey (1967) was able to show that in the irradiation of phage T7 with accelerated argon ions, single strand breaks are only 2.3 times as frequent as double strand breaks, whereas for gamma irradiation this value is 5—6.

After *irradiation of DNA in dilute aqueous solutions* the number of single strand breaks per nucleotide (Fig. 74) is proportional to the dose:

$$B_1 = k \cdot D, \qquad (10.4)$$

were k is the probability of a break per nucleotide and per rad; at a DNA concentration of 0.2 mg/ml the value obtained is $k = 4.15 \cdot 10^{-7}$ rad^{-1} (Hagen, 1967). The formation of double strand breaks, in contrast, follows a different kinetic. As can be seen from Fig. 74, their frequency increases with the square of the dose; this result is also obtained by viscosity measurements (Cox *et al.*, 1955). This dependence shows that the two strands are not broken simultaneously by the attack of the water radicals, which seems reasonable. A break in the double helix is produced only if two single

145

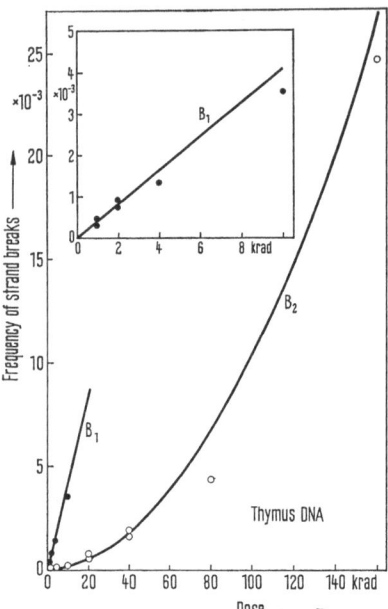

Fig. 74. Single and double strand breaks in calf thymus DNA after irradiation in aqueous solution (0.2 mg/ml). B_1 = frequency of single strand breaks per nucleotide. B_2 = frequency of double strand breaks per nucleotide pair. (Hagen, 1967)

breaks are either on exactly or approximately opposite locations. In the latter case, the hydrogen bonds between these two locations have to be opened for the double break to be detectable. The number of double strand breaks per nucleotide pair in solution, taking equation (10.4) into account is given by:

$$B_2 = (\beta + B_1)^2 \cdot n = (\beta + kD)^2 \cdot n, \qquad (10.5)$$

where $(n-1)/2$ is the maximum number of nucleotide pairs between two single strand breaks leading to a double break and β is the number of single strand breaks present at the start of irradiation. In unirradiated samples, β is usually so small (e. g. 10^{-4}; Hagen, 1967), that it can be neglected at high doses. The curve shown in Fig. 74 has been fitted to the points using equation (10.5) with $n = 7$. This indicates that a double strand break occurs if in the presence of a single break on one strand, a subsequent break occurs on the complementary strand exactly opposite, or not more than three nucleotide pairs distant, from the first break.

If the G-values for the occurrence of single strand breaks in irradiated DNA, as measured by various authors (see the tables by Collyns et al., 1965; Ginoza, 1967; as well as Weinert and Hagen, 1968) are compared, it is found in aqueous solutions that most results lie between 0.3 and 0.8, and

146

for DNA irradiated in the dry state, in cells or as a nucleoprotein-gel, between 0.3 and 0.7; i. e. approximately the same amount of energy is required for a single strand break in the dry state and in solution. Since the double breaks in solution increase with the square of the dose, no G-values (in the normal sense) can be given. Initially, only single strand breaks are produced, but with increasing doses the proportion of double to single breaks gradually increases. It is, therefore, to be expected that in an aqueous system with a test-response that occurs only after a double strand break, the dose-response curves will possess a shoulder (for example the inactivation of phage T7 in buffer; see Fig. 92). In dry DNA, the number of double breaks increases linearly with dose, as has already been mentioned. The G-values, as measured by various authors, lie between 0.1 and 0.15; they are only marginally larger in the presence of oxygen than under nitrogen or in vacuo.

In conclusion, it should be pointed out that after UV irradiation of DNA, breaks in the polynucleotide strands are so rare that they are of no biological importance, since even after a UV dose which inactivated 99⁰/o of phage T7, no breaks could be found in the DNA strands (Freifelder and Davison, 1963).

10.5. Intermolecular Cross-Linking

In the *irradiation of dry DNA*, links between individual DNA molecules may occur, as well as breaks in the polynucleotide chains. This is clearly demonstrated in the sedimentation diagrams of Fig. 75. The sedimentation distribution of the control shows the inhomogeneity in the molecular weights of dried thymus DNA, since molecules of uniform size should all migrate at the same speed and give a sharp line. After irradiation, the maximum of the distribution has shifted to a smaller S-value (S = Svedberg, unit of the sedimentation coefficient, $1 S = 10^{-13}$ sec). The main fraction of the molecules therefore migrate more slowly, i. e. the molecular weight is smaller than the average of the control. This is additional evidence for the occurence of double strand breaks in the irradiation of dry DNA. As well as this shift of the maximum of the distribution to the right, an increase in the amount of DNA in the region 30—55 S is observed in the sedimentation diagram. This fraction consists of molecules linked together, and having molecular weights of 20 to $100 \cdot 10^6$ Dalton. This fraction of the total material is much smaller after irradiation under oxygen than after irradiation under nitrogen. The G-value for cross-linking is 0.16 in the presence of oxygen and 0.37 in vacuo (Hagen and Wellstein, 1965). Similar results are obtained with irradiated proteins, in which the presence of oxygen partially inhibits the process of dimerization, so that unfolded monomers are formed (see Chapter 9.5). These results also show that the

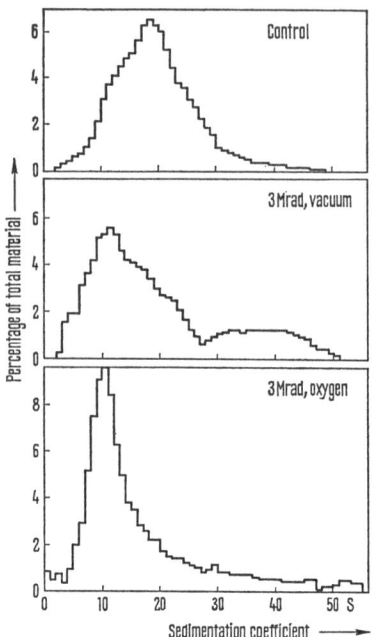

Fig. 75. Sedimentation pattern of calf thymus DNA after irradiation in the dry state. (Hagen and Wellstein, 1965)

formation of cross-links in irradiated DNA involves free radicals, which can react with oxygen to form peroxy-radicals, whereby a part of the cross-linking (equation 9.8) is inhibited.

Upon irradiation of *DNA in aqueous solutions,* the situation is much more complex than in the dry state. The rate of cross-linking is influenced by numerous parameters, including the size, conformation and concentration of the macromolecules, ionic strength of the solvent, polar effects, etc. The results of investigations on synthetic macromolecules (cf. Henglein and Schnabel, 1966) showed that there is an optimal concentration for the production of cross-links. At higher concentrations, the rate of cross-linking per unit dose decreases, and gradually approaches the value obtained in the dry state. As the concentration decreases, the frequency of cross-linking also decreases, until below a critical concentration no cross-linking is detected. This critical concentration decreases as the molecular weight of the irradiated molecule increases. If irradiation is carried out in the intermediate concentration range, it is therefore possible that cross-links will be formed initially, but that with increasing dose the degradation may reduce the molecular weight of the fragments, thereby increasing the critical concentration to a level above that of the DNA fragments. Thus, in these conditions, no cross-linking reactions will be observed at higher doses.

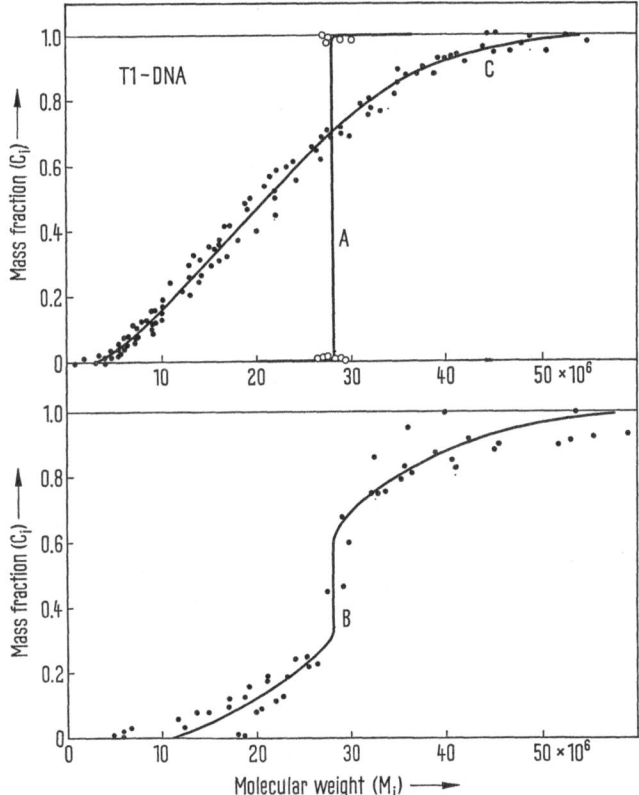

Fig. 76. Distribution of molecular weights of DNA from unirradiated and irradiated T1-bacteriophage. C_i is the proportion of molecules with a weight smaller than M_i. Curve A: unirradiated DNA. Curve B: 1 krad ^{60}Co γ-radiation. Curve C: 4 krad ^{60}Co γ-radiation. Irradiation in solution; concentration: 0.2 mg/ml. (Coquerelle et al., 1969)

Unfortunately the rate of cross-linking of DNA irradiated in aqueous solution has not yet been examined as a function of the various parameters. It can only be noted that radiation also causes cross-linking of molecules in solution. Coquerelle et al. (1969) irradiated DNA solutions of bacteriophage T1 (0.2 mg/ml) and determined the distribution of molecular weights by measurements of viscosity and sedimentation rate at various concentrations. In the centrifugation of unirradiated phage-DNA, the gradient (i. e. the transition from the solvent to solution) is quite sharp (Fig. 76, perpendicular line); this shows the uniformity in DNA preparations that can be obtained by careful isolation from bacteriophages. After a dose of 1 krad, some of the DNA remains unchanged (Fig. 76, perpendicular portion of curve B), approximately one-third is degraded, and one-third has increased in size. After 4 krad, there are only a few molecules of the original size; about

70% of the irradiated material has a lower molecular weight than the unirradiated DNA, and approximately 30% has a molecular weight larger than the control, due to cross-linking. Above 8 krad, degradation predominates over cross-linking. According to Charlesby (1960) the ratio of breaks to cross-links can be calculated if the weight-average molecular weight M_w and the number-average molecular weight M_n of the distribution is known. In the experiment described here, one cross-link is formed for each 4.7 double breaks (Coquerelle et al., 1969). It is to be expected that this ratio will depend greatly on the experimental conditions.

10.6. Rupture of Hydrogen Bonds

The number of ruptured hydrogen bonds in an irradiated DNA molecule can, for example, be measured by electrometric titration of the amount of acid required to denature the DNA (cf. Cox et al., 1958). The degree of denaturation of DNA can also be determined using the "hypochromicity" (sometimes called hyperchromicity or hyperchrome effect). This results from the fact that the optical absorption of native double-stranded DNA is smaller than would be predicted from the number of nucleotides, i. e. the extinction coefficient of the bases in a single strand or in free nucleotides is about 30% higher than in an intact double-stranded DNA molecule linked by hydrogen bonds. Dispersion, polar effects of the solvent, influence of exciton interactions on oscillator strength, and other processes have been discussed as reasons for this change in optical absorption (cf. Zimm and Kallenbach, 1962; Scholes, 1963).

If the number of hydrogen bonds decreases as a result of irradiation, the optical absorption should increase. This is observed in most cases. However, this increase is not, in itself, a measure of the number of ruptured hydrogen bonds, since a fraction of the bases is also simultaneously decomposed thereby decreasing the absorption. These two opposing processes can be distinguished by measurement of the extinction in neutral and acidic solutions. As shown in Fig. 77, the extinction due to the hypochromicity increases by 30—40% on acid denaturation of unirradiated double-stranded DNA (pH less than 3.5). In contrast, the absorption of the irradiated DNA is higher in neutral solution and lower in acidic solution, than in the unirradiated control. The decrease in the absorption below pH 3.5 is due purely to the destruction of the bases, while the increase in neutral solution is due to the hypochromicity superimposed on the decrease caused by the base decomposition. If the extinction at pH 2.5 is plotted as a function of dose, then a linear decrease is observed, except at very high doses (Fig. 78). The same curve is obtained if the DNA is hydrolyzed with formic acid after irradiation, thereby releasing all the bases. This shows that the extinction in the acidic region is a measure of the number of undamaged bases, provided that the irradiation products generated do not absorb at

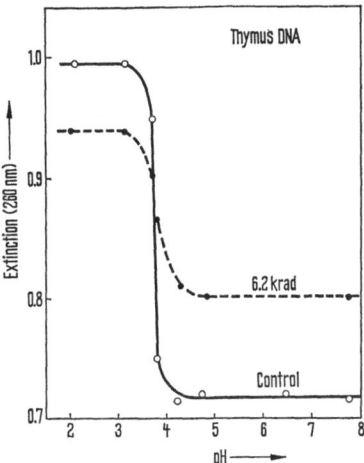

Fig. 77. Influence of pH on the optical absorption of unirradiated and irradiated calf thymus DNA. Irradiation in solution (0.006%) with 6.2 krad ^{60}Co γ-radiation, measured at a concentration of 0.005%. (Collyns *et al.*, 1965)

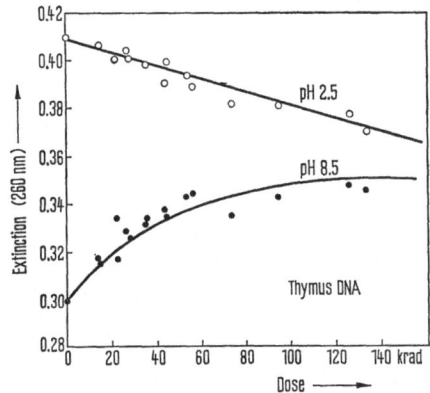

Fig. 78. Changes in the optical absorption of calf thymus DNA at pH 2.5 and 8.5 after irradiation in aerobic solution (0.1%) with ^{60}Co γ-radiation. (Collyns *et al.*, 1965)

260 nm; this assumption was shown to be correct (cf. Weiss, 1964). When all the hydrogen bonds are destroyed at high doses, the absorption values coincide in the neutral and acidic regions, and decrease with increasing radiation-induced base decomposition.

Fig. 78 also shows that the breakage of hydrogen bonds cannot be correlated with the destruction of bases. At 140 krad, for example, more than 90% of these bases are still undamaged, while the hypochromicity has decreased to 20% of its original value. However, the number of broken hydrogen bonds cannot be determined directly from this, since there is not a linear relationship between the number of intact hydrogen bonds and the change in absorption caused by denaturation. The exact relationship is not known, and semi-empirical equations therefore have to be used (Applequist, 1961). The curves presented in Fig. 78, determined by Collyns *et al.* (1965), give a yield of $G = 6.6$ for the rupture of one hydrogen bond. The G-values reported by various authors lie in the range between 2.7 and 60 (see the tables given by Collyns *et al.*, 1965, and Ginoza, 1967). The reason for this large variation may be that often the non-linear relationship between hypochromicity and number of hydrogen bonds ruptured was not taken into account.

According to Scholes *et al.* (1960), about 14—15 hydrogen bonds are ruptured per single strand break at low doses, and according to Peacocke and Preston (1959) about 16. This means that at the site of a single strand break, there are on average three to four nucleotide pairs no longer linked by hydrogen bonds. This value agrees with the previously mentioned observations of Hagen (1967) that single breaks on each DNA strand less than three nucleotides apart give rise to a double break (see Chapter 10.4). The relatively large number of hydrogen bonds opened per single strand break can be explained by the penetration of water molecules into the double helix following the breakage of a single strand. As a result, the hydrogen bonds between complementary bases are to some extent replaced by bonds between the individual bases and water molecules, so that the polynucleotide strands are effectively "unzipped" (Scholes *et al.*, 1960). Of course, this process must be restricted in some way, otherwise there would be no double-stranded DNA left in an aqueous solution. On the basis of thermodynamical considerations it can be estimated that in a system of two chains twined around each other, the unwinding gives an initial gain in entropy per unwound bond. This has a maximum value and finally reaches a limit where the untwisted region of the chain reaches a certain "equilibrium length". Temperley (1959) calculated that in DNA the unwound section contains 15—20 hydrogen bonds.

The influence of radiation on the stability of hydrogen bonds can also be detected by *melting curves*. Denaturation occurs when double-stranded DNA is heated in solution, resulting in an increase in the extinction coefficient at temperatures above 65° C, which reaches a constant level again at about 90° C. The temperature at which the increase in extinction reaches 50% of its optimal value is known as the "melting point". The melting-temperature decreases with increasing radiation dose (see Hagen and Wild, 1964), which is a further indication of the radiation-induced labilization of the structure of the double helix. This decrease in the melting point is observed at doses considerably lower than those causing the destruction of the bases. It may, therefore, be assumed that this effect is also a function of the frequency of single breaks, since these reduce the longitudinal stability of the molecule and provide additional points for the unwinding of the helix.

The *production of denatured regions* is the last criterion of changes in hydrogen bonds after irradiation that will be discussed; it can be detected chromatographically using columns of methylated albumin on kieselguhr (MAK). The fraction of elutable material decreases exponentially with dose after irradiation of DNA in aqueous solution (Ullrich and Hagen, 1968). If the DNA is degraded ultrasonically to one quarter of its molecular weight after irradiation, the fraction of elutable material increases by a factor of 4, demonstrating that this procedure divides the molecule into damaged and undamaged sections. It was shown by Ullrich and Hagen (1968), that the DNA molecules bound to the MAK-column contain "denatured regions". These are small regions, such as those produced thermally or by shearing forces, in which the two nucleotide strands are no longer linked by hydrogen bonds. These lesions cannot be renatured. Increasing the number of single strand breaks by treatment with DNase does not cause a parallel increase in the number of denatured zones. Thus, these lesions must be distinguished from the previously discussed mechanism according to which most of the broken hydrogen bonds are derived from single strand breaks. In aqueous solution, irradiation produces about 4 times as many single strand breaks as denatured regions. Further experiments are necessary before more can be said about the nature and formation of these denatured zones. Although these examples of the use of physico-chemical methods in the investigation of changes in irradiated DNA represent but the beginning of such applications, a series of interesting results have already been obtained, as has been seen. The future development and application of these methods should make further significant contributions to the elucidation of the molecular mechanisms of the action of radiation.

References

Alexander, P., Lett, J. T., Kopp, P., Itzhaki, R.: Radiat. Res. 14, 363 (1961).
Applequist, J.: J. Am. Chem. Soc. 83, 3158 (1961).
Beukers, R., Berends, W.: Biochim. biophys. Acta (Amst.) 41, 550 (1960).
Charlesby, A.: Atomic radiation and polymers. Oxford: Pergamon Press 1960.
Collyns, B., Okada, S., Scholes, G., Weiss, J. J., Wheeler, C. M.: Radiat. Res. 25, 526 (1965).
Cook, J. B., Elliott, J. P., Wyard, S. J.: Mol. Phys. 13, 49 (1967).
Coquerelle, T., Bohne, L., Hagen, U., Merkwitz, J.: Z. Naturforsch. 24 b, 885 (1969).
Cox, R. A., Overend, W. G., Peacocke, A. R., Wilson, S.: Nature 176, 919 (1955).
— — — — Proc. roy. Soc. (Lond.) B 149, 511 (1958).
Daniels, M., Scholes, G., Weiss, J. J., Wheeler, C. M.: J. chem. Soc. (Lond.) 1957, 226 (1957).
Dertinger, H.: Z. Naturforsch. 22 b, 1261 (1967a).
— Z. Naturforsch. 22 b, 1266 (1967b).

Dewey, D. L.: Int. J. Radiat. Biol. **12**, 497 (1967).
Dorlet, C., van de Vorst, E., Bertinchamps, A. J.: Nature **194**, 767 (1962).
Ehrenberg, A., Ehrenberg, L., Löfroth, G.: Nature **200**, 376 (1963).
Eigner, J., Doty, P.: J. molec. Biol. **12**, 549 (1965).
Freifelder, D.: Proc. nat. Acad. (Wash.) **54**, 128 (1965).
— Davison, P. F.: Biophys. J. **3**, 97 (1963).
Ginoza, W.: Ann. Rev. Microbiol. **21**, 325 (1967).
Hagen, U.: Biochim. biophys. Acta (Amst.) **134**, 45 (1967).
— In: Experimental methods in molecular biology. Ed.: C. Nicolau. London: John Wiley & Sons, in press, 1970.
— Wellstein, H.: Strahlentherapie **128**, 565 (1965).
— Wild, R.: Strahlentherapie **124**, 275 (1964).
Heller, H. C., Cole, T.: Proc. nat. Acad. Sci. (Wash.) **54**, 1486 (1965).
Hems, G.: Nature **186**, 710 (1960).
Henglein, A., Schnabel, W.: In: Current topics in radiation research, Vol. II. Eds.: M. Ebert and A. Howard. Amsterdam: North-Holland Publ. Co. 1966, p. 1.
Herak, J. N., Gordy, W.: Proc. nat. Acad. Sci. (Wash.) **54**, 1287 (1965).
Hershey, A. D., Goldberg, E., Burgi, E., Ingraham, L.: J. molec. Biol. **6**, 230 (1963).
Hüttermann, J., Müller, A.: Radiat. Res. **38**, 248 (1969).
Kanazir, D. T.: Progr. Nucleic Acid. Res. **9**, 117 (1969).
Keck, K.: Z. Naturforsch. **23 b**, 1034 (1968).
Köhnlein, W.: Strahlentherapie **122**, 437 (1963).
Latarjet, R., Ekert, B., Demerseman, P.: Radiat. Res. Suppl. **3**, 247 (1963).
Lett, J. T., Stacey, K. A., Alexander, P.: Radiat. Res. **14**, 349 (1961).
Lichter, J. D., Gordy, W.: Proc. nat. Acad. Sci. (Wash.) **60**, 450 (1968).
McCargo, M.: Cited by Weiss 1964.
Müller, A.: Int. J. Radiat. Biol. **6**, 137 (1963).
— Int. J. Radiat. Biol. **8**, 131 (1964).
— Progr. Biophys. **17**, 99 (1967).
Peacocke, A. R., Preston, B. N.: Cited by Scholes et al., 1960.
Pershan, P. S., Shulman, R. G., Wyluda, B. J., Eisinger, J.: Physics **1**, 163 (1964).
Pruden, B., Snipes, W., Gordy, W.: Proc. nat. Acad. Sci. (Wash.) **53**, 917 (1965).
Salovey, R., Shulman, R. G., Walsh, W. M.: J. Chem. Phys. **30**, 839 (1963).
Schmidt, J., Snipes, W.: Int. J. Radiat. Biol. **13**, 101 (1967).
Scholes, G.: Progr. Biophysics **13**, 59 (1963).
— Ward, J. F., Weiss, J. J.: J. molec. Biol. **2**, 379 (1960).
— Weiss, J. J.: J. exp. Cell Res. Suppl. **2**, 219 (1952).
— — Biochem. J. **56**, 65 (1954).
Setlow, R. B., Setlow, J. K.: Proc. nat. Acad. Sci. (Wash.) **48**, 1250 (1962).
Smith, K. C., Hanawalt, P. C.: Molecular photobiology. New York: Academic Press 1969.
Temperley, H. N. V.: Trans. Faraday Soc. **55**, 515 (1959).
Ullrich, M., Hagen, U.: Z. Naturforsch. **23 b**, 1176 (1968).
Wacker, A., Dellweg, H., Jacherts, D.: J. molec. Biol. **4**, 410 (1962).
— Lochmann, E. R.: Z. Naturforsch. **17 b**, 351 (1962).
Weinblum, D., Johns, H. E.: Biochim. biophys. Acta (Amst.) **114**, 450 (1966).
Weinert, H., Hagen, U.: Strahlentherapie **136**, 204 (1968).
Weiss, J. J.: Progr. Nucleic Acid Res. **3**, 103 (1964).
Zimm, B. H., Kallenbach, N. R.: Ann. Rev. Phys. Chem. **13**, 171 (1962).
Zimmer, K. G., Müller, A.: In: Current topics in radiation research, Vol. I. Eds.: M. Ebert and A. Howard. Amsterdam: North-Holland Publ. Co. 1965, p. 1.

Chapter 11. Inactivation of Nucleic Acid Functions

11.1. Functions of Nucleic Acids

The nucleic acids fulfil important functions within the complex biochemical processes which are the basis of what we know as life. For example, the total genetic information of an individual is contained in its nucleic acids; this information is carried in most organisms by double-stranded DNA, but in some viruses the genome consists of a single strand of DNA or RNA, and in rare cases of double-stranded RNA. Duplication of the genetic material must occur to enable the genetic information to be transferred from one generation to the next (for example, in cell division). According to the Watson-Crick mechanism for the semi-conservative *replication* of double-stranded DNA, the original strands separate while the complementary strands are synthesized. This process can also occur in vitro; such a system must contain DNA polymerase and deoxyribonucleoside-triphosphate, as well as a DNA template, also known as "primer" DNA (cf. Kornberg, 1961).

In protein biosynthesis, the information carried by DNA is transferred to the messenger RNA (mRNA). The mechanism of this *transcription* is comparable with that of DNA synthesis, as it also requires a polymerase enzyme and a reservoir of ribonucleoside-triphosphates. However, not all of the DNA strand is copied as a whole (as in replication) but sections of the DNA, which seem to have "start-and-stop" sites for the RNA polymerase. These sections have different lengths, and the corresponding molecular weights of mRNA are rather heterogeneous, lying in the range 100 000 to 500 000 or even higher. The transcription of genetic information from DNA to the smaller messenger RNA has the advantage that this latter information-carrying molecule has a greater mobility. The mRNA then migrates to the ribosomes, where the proteins are synthesized. The conversion of the nucleotide sequence of mRNA into an amino acid sequence is called *translation*. It requires another type of nucleic acid, known as transfer RNA (tRNA). Each amino acid is bound to a specific tRNA, which recognizes on the mRNA the codons characteristic for "its" amino acid, so that the corresponding amino acids are incorporated into the protein in an order determined by the DNA.

The total information carried in the DNA of a cell is not being read continuously and used for the synthesis of proteins, either in the specialized cells of a higher organism or in primitive single-cell organisms. Instead, there is a pronounced *regulation* of the activity of the individual sections of DNA. If, for example, E. coli bacteria using glucose as their energy source are grown in the absence of lactose, then only traces of the enzymes galactoside-permease and β-galactosidase, which are required for the metabolism of lactose, are formed. After transfer to a medium containing lactose but no glucose, the bacteria produce both enzymes in quantities up to 10,000 times greater than when grown in glucose; this process is referred to as "enzyme induction". As well as these universal nucleic acid functions, there are some other functions specific to viruses and bacteria, such as infectivity and transformation, which will be discussed later. More details of the processes and functions that have been sketched here are presented, for example, in the book of Watson (1965).

Various functions of nucleic acids can be inhibited or destroyed by irradiation. It is already apparent from the number and variety of aspects of the processes mentioned that the term "inactivation" can have no simple meaning for nucleic acids. In contrast to enzymes which, at least with respect to their inactivation kinetics, respond to irradiation in a consistent manner (see Chapter 9.2), the inactivation of almost every nucleic acid function follows different kinetics. Since none of the inactivation mechanisms considered here are completely understood, this chapter will concentrate on the discussion of the different inactivation kinetics, and on the special aspects of the corresponding radiation damage which can be derived from them. In discussing individual examples, other observations will be mentioned in so far as they will contribute to an understanding of the inactivation processes.

11.2. Infectivity

The term "infectivity" refers to the ability of a virus to multiply in suitable host cells. If the host cell is a bacterium, the process of infection is initiated by injection of the viral DNA (or RNA) into the bacterium; this will be discussed in more detail in Chapter 12.1. After about twenty minutes, the bacterium is lysed, i. e. the bacterial cell wall is ruptured, releasing about 100 to 200 new phages. The reproduction process shows clearly that the information required for the formation of a new phage, including its protein coat, is carried in the phage DNA. This suggested that spheroplasts (bacteria treated with lysozyme to remove a portion of their cell wall) could be infected with DNA isolated from bacteriophage, thereby initiating the formation of complete phages. This system, using "infectious" DNA, functions particularly well with the single-stranded circular DNA from the bacteriophage ΦX 174 (Guthrie and Sinsheimer, 1963), but it also works

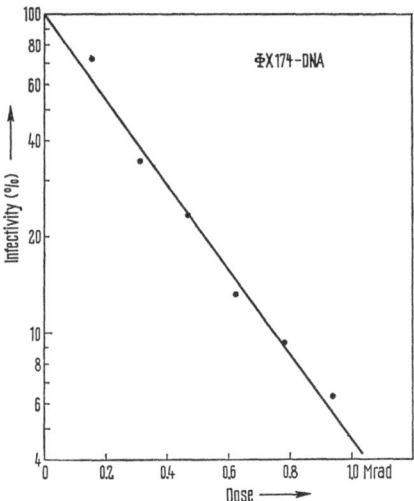

Fig. 79. Inactivation of freeze-dried infectious ΦX 174-DNA in vacuo by ^{60}Co γ-irradiation. (Jung, 1968)

with phage T1 and certain other viruses. The number of phages released from infected spheroplasts is proportional to the concentration of infectious DNA over several orders of magnitude, and can be determined by standard procedures (see Chapter 12.1).

The irradiation of pure, freeze-dried ΦX 174-DNA in vacuo, with ^{60}Co γ-radiation, gives an exponential dose-response curve (Fig. 79). The molecular weight of the target can be calculated from the D_{37} of 320 krad, using equation (5.5); it amounts to $1.8 \cdot 10^6$ Dalton, which is in good agreement with the molecular weight of ΦX-DNA of $1.7 \cdot 10^6$ Dalton (Sinsheimer, 1959). This means that a single interaction between radiation and ΦX-DNA, with a mean energy transfer of 60 eV, is sufficient to inhibit the formation of intact phages, i. e. to destroy the infectivity.

According to Lytle and Ginoza (1969) about 25 per cent of the ΦX-DNA molecules inactivated by direct radiation action carry a break in their polynucleotide chain. From this value and from the energy deposition of 60 eV per inactivated DNA molecule (corresponding to a G-value of 1.65), the G-value for the production of a strand break in ΦX-DNA is calculated to be 0.4. This result agrees well with the G-value for the generation of single strand breaks in double-stranded DNA, determined by ultracentrifugation (cf. Chapter 10.4). The fact that the infectivity is lost when circular ΦX-DNA is converted to a linear structure by a strand break has also been shown directly using DNase (Fiers and Sinsheimer, 1962). The remaining 75% of radiation inactivation is probably caused by damage to the bases. It should be expected that almost all base changes lead to inactivation, since in solution 20—40% of radical attacks are on the sugar, and 60—80% on the bases (Chapter 10.3). However, this problem has not yet been resolved, as there are certain situations in which up to 10 damaged bases per inactivated DNA molecule have been found (Blok, 1967; Jung et al., 1969).

11.3. Transformation

Transforming ability provided some of the classical proofs of the statement that DNA is the carrier of genetic information. It was discovered by Griffith (1928) using pneumococcus, and has since been observed for certain other kinds of bacteria (for example Haemophilus influenza and Bacillus subtilis), but not for all (see review article of Ravin, 1961). In the transformation process, a "competent" bacterium takes up specific DNA fragments from the milieu, and incorporates their genetic information into the bacterial chromosome, so that it consequently becomes heritable. To demonstrate transformation, DNA is extracted from bacteria having a particular genetic property ("marker") such as resistance to streptomycin. This transforming DNA is subsequently incubated with mutants of the same bacterial strain but without that marker. Colony formation is then studied, using a medium containing a specific streptomycin concentration, on which only bacteria that have acquired the streptomycin resistance by transformation are able to form colonies. As well as the diverse antibiotic resistant mutants, there are many others suitable for transformation experiments, such as those which lack the ability to synthesize a certain substance and, therefore, require that substance in their nutrient medium for growth (auxotrophic mutants).

In radiation biological experiments, the transforming DNA (formerly referred to as "transforming principle") is irradiated, and the number of transformed cells in a population is determined as a function of dose. Since ionizing radiation and UV light inactivate transforming DNA in different ways, their effects will be discussed separately.

a) Ionizing Radiation

If spores or vegetative cells of B. subtilis are irradiated with γ-rays or electrons, and their DNA is then extracted and incubated with indole-requiring cells, then the frequency of transformation of this indole marker decreases exponentially with dose (see Fig. 80). The DNA in the vegetative cells ($D_{37} = 1.56$ Mrad) is 4 times more sensitive to radiation than that in dry anaerobic irradiated spores ($D_{37} = 6.3$ Mrad). The higher sensitivity in wet cells may be attributable to water radicals (see Chapter 6.3); however, the possibility that oxygen may contribute to the inactivation in vegetative cells cannot be ruled out. On the other hand, D_{37}-values of 27 and 115 krad are obtained if a colony-forming test is carried out with irradiated cells and

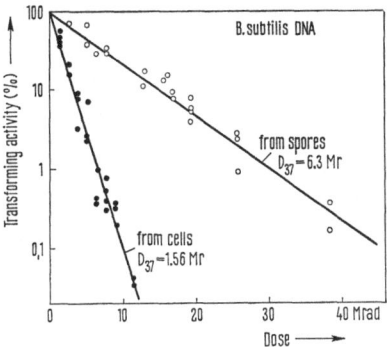

Fig. 80. Inactivation of transforming DNA of Bacillus subtilis by irradiation of vegetative cells and dried spores with 1 MeV electrons. (Tanooka and Hutchinson, 1965)

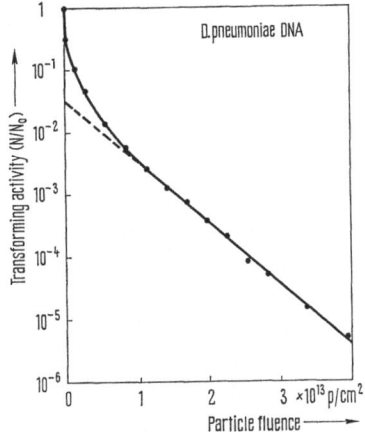

Fig. 81. Inactivation of dry transforming DNA of Diplococcus pneumoniae by 10 MeV protons. (Guild and Defilippes, 1957)

spores, respectively (Tanooka and Hutchinson, 1965). In this case, the radiation sensitivities again differ by a factor of 4; the overall sensitivity of transformation is, however, 60 times lower than that of the colony-forming ability. This indicates that the radiation sensitive target for the inhibition of transformation may be only a small segment of the DNA, approximately the size of the marker. A target molecular weight of about 100,000 for the indole marker of B. subtilis is obtained from the D_{37} of 6.3 Mrad (cf. equation 5.5).

However, exponential dose-response curves are the exception rather than the rule in transformation experiments. Figure 81 shows the transforming activity of isolated DNA irradiated in the dry state with 10 MeV protons; it decreases rapidly at first, and approaches exponential dependence at high doses. Similar responses are found using other markers and radiations of different qualities (Guild and Defilippes, 1957; Lerman and Tolmach, 1959). The exponential portion of such transformation curves leads, according to equation (5.5), to MW_T-values of about $2 \cdot 10^5$ Dalton, which is also approximately the size of a marker (Guild and Defilippes, 1957; Hutchinson, 1962). Because of the importance of radiation-induced double strand breaks, it must be considered whether this approach should not be replaced by one which expresses the dependence of transformation probability on the length of the transforming DNA molecule (Cato and Guild, 1968).

b) UV Light

The action of ultraviolet light produces a different situation. For example, it is interesting that there are markers which are equally sensitive to the action of X-rays, heat, chemicals and DNase, but whose UV-sensitivities differ by factors of 5 to 10 (Lerman and Tolmach, 1959). It is also found that the slope of the UV transformation curve decreases continuously with increasing dose, i. e. it does not, as in the case of ionizing radiation, approach exponential dependence asymptotically. Rupert and Goodgal (1960) have shown that the UV curves can be represented by the formula:

$$N/N_0 = 1/(1 + kD)^2 . \qquad (11.1)$$

This implies that a straight line passing through 1 is obtained when $\sqrt{N_0/N}$ is plotted against the UV dose. This is confirmed experimentally as can be seen, for example, from Fig. 82 which shows the transformation of cathomycin and streptomycin resistance in Haemophilus influenzae. This type of plot has the advantage that the difference between the sensitivities of the markers, which depend on the dose in a logarithmic plot, can be specified quantitatively by the factor k in equation (11.1).

Theoretical Approach. The problem now arises of constructing a model to describe this unusual type of dose-dependence of radiation action, which has not been encountered in the previous ten chapters. This is rendered even more necessary by the similar dose-response curves obtained for the actions of heat, nitrogen mustard, hydrazine, hydroxylamine, dimethylsulphate, nitrous acid and DNase on transforming DNA (Lerman and Tolmach, 1959; Bresler et al., 1967). A theoretical expression for the solution of this problem has been proposed by Bresler and colleagues (1967). The basic idea of their theory is that they consider the whole DNA molecule to be the primary target, and not the individual marker. The decrease in radiation sensitivity with increasing dose, which is shown by the bend in the dose-response curve, is explained by the shortening of the effective recombination length, resulting from a random distribution of radiation lesions. This assumption is justified in as much as in the investigation of two adjacent markers, the number of simultaneously transformed cells is approximately equal to the number

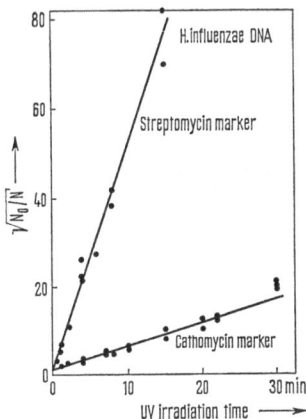

Fig. 82. Inactivation of the streptomycin and cathomycin markers of transforming DNA from Haemophilus influenzae by irradiation with 254 nm UV light at an intensity of 1650 ergs/cm²min. (Rupert and Goodgal, 1960)

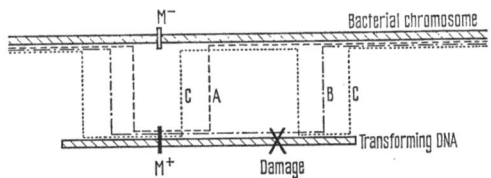

Fig. 83. Scheme of the genetic recombination between a transforming DNA molecule carrying the marker M^+ and a bacterial chromosome deficient in this marker (M^-). (Bresler et al., 1967)

which have undergone a single transformation. The scheme for recombination between the transforming DNA molecule and the bacterial chromosome is given in Fig. 83. In the unirradiated state, every recombination involving the marker M leads to a successful transformation (Fig. 83; A and C). No transformation occurs if the recombination includes a radiation lesion, since this forms a type of lethal mutation (B). Thus, if the number of lesions in a DNA molecule increases with increasing dose, then the number of recombinations leading to viable transformants decreases.

If recombination is a rare event, the target is represented as a distribution of distances between the marker and the nearest lesion. The probability of recombination is then a function of this distribution. (Bresler et al., 1964, based their calculations on this model). If, however, recombinations occur relatively frequently, as they do in the bacteria discussed here, then transformation also occurs when the recombinations are on both sides of the lesion (Fig. 83, example C). In this case, the recombination probability (W) is described as a function of the recombination length l by the expression

161

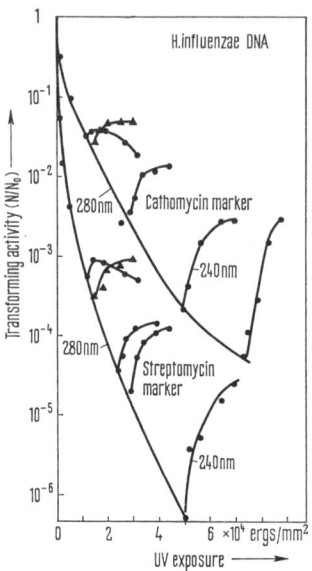

Fig. 84. Inactivation of the streptomycin and cathomycin markers of transforming DNA from Haemophilus influenzae by UV light of 280 nm wavelength, and "reactivation" by a subsequent irradiation at 240 nm. (Setlow and Setlow, 1962)

of Haldane (1919) as:

$$W(l) = \frac{1}{2}(1 - e^{-2\omega l}). \tag{11.2}$$

The following approximate expression for the transformation rate under these conditions is obtained after lengthy calculation (see Bresler *et al.*, 1967):

$$N/N_0 = 1 \Big/ \Big(1 + \frac{z}{2\omega L}\Big)^2, \tag{11.3}$$

where L is the mean length of the transforming DNA molecule, z is the mean number of lesions per molecule and $1/\omega$ the genetic unit of length. Since z is proportional to the dose, this formula corresponds to the semi-empirical expression (11.1) postulated by Rupert and Goodgal (1960).

If the theory of Bresler *et al.* is applied to coupled markers sufficiently far apart on the DNA molecule, then it is to be expected that the rate of transformation as a function of dose will be given by:

$$N/N_0 = 1/(1 + kD)^4. \tag{11.4}$$

This has been demonstrated experimentally by Rupert (1968), using two pairs of coupled markers with Haemophilus influenzae, where each individual marker lost its transforming ability following equation (11.1).

162

Little is known about the *type of damage* that leads to the loss of transforming activity, particularly in the case of ionizing radiation. In the case of UV light it is known that a proportion of the inactivation of the transforming DNA is due to the formation of thymine dimers. As already mentioned in Chapter 10.3, the yield of dimers is greatest at 280 nm, while at 240 nm the splitting of the dimer is favoured. The inactivation of the two markers for streptomycin and cathomycin resistance of H. influenzae at 280 nm follows the usual dose-response curve (Fig. 84). If a second irradiation at 240 nm is then carried out, the transformation rate increases again as a result of the splitting of some of the thymine dimers. Nevertheless, not all of the inactivation can be attributed to UV-induced thymine dimers (see Chapter 10.3).

11.4. Priming Activity of DNA

In Chapter 11.1, it was mentioned that the synthesis of new DNA (replication) and of messenger RNA (transcription) can be carried out in vitro, using DNA as a primer. To measure the priming activity, irradiated DNA is incubated with DNA or RNA polymerase together with the corresponding nucleoside triphosphates. The triphosphates are usually labelled with ^{3}H, ^{14}C or ^{32}P, and their uptake into the acid-insoluble fraction of the nucleic acids is studied. Such tests have been carried out by Harrington (1964) using both the DNA and the RNA polymerase systems. In each case, dose-response curves with an upward curvature were obtained (Fig. 85) after exposing the primer DNA in dilute solution to ionizing radiation, leading to a response similar to that obtained in transformation experiments. It is worth noting that the priming activity of irradiated thymus DNA is about 40 times more sensitive for DNA synthesis than for RNA synthesis. The different sensitivities of these two processes can be explained by taking into account the fact that the whole molecule is read in DNA synthesis, while in RNA synthesis only limited sections are copied in one piece; thus, the target for the DNA polymerase system is considerably larger than that for the RNA system.

The bend in the dose-response curve could possibly be attributed to three mechanisms:

(1) Targets of different sizes are present, corresponding to the observed inhomogeneity of the molecular weight distribution of the synthesized mRNA (see Chapter 11.1). Since at low doses the larger DNA sections are predominantly inactivated, the dose-response curve falls off rapidly initially, but more slowly at higher doses (Fig. 6).

(2) As a result of the irradiation, new binding sites for the polymerase are formed which do not, however, serve as starting points for RNA synthesis (Weiss and Wheeler, 1967). It is assumed that there are *n* natural

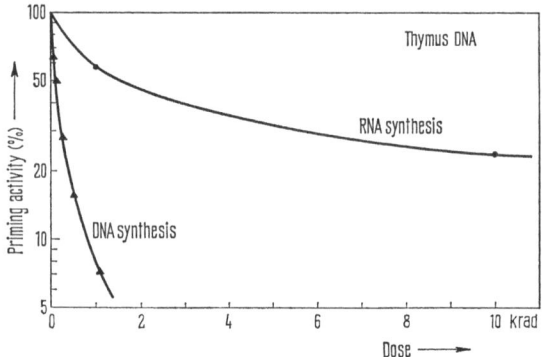

Fig. 85. Inactivation of the priming activity of calf thymus DNA for DNA and RNA synthesis, respectively. Irradiation in phosphate buffer (0.05 mg/ml) with ^{60}Co γ-rays. (Harrington, 1964)

starting points and r new binding sites formed during irradiation. Since mRNA synthesis commences only from the original starting points, the synthesis after irradiation is reduced from A_0 (the amount of mRNA in the unirradiated control) to A, depending on the ratio of active to total binding sites for the enzyme on the DNA molecule:

$$A/A_0 = n/(n-r). \tag{11.5}$$

Since r can be considered as being proportional to dose, the expression can be rewritten using a suitable constant k:

$$A/A_0 = 1/(1+kD). \tag{11.6}$$

Therefore, if A_0/A is plotted against dose, a straight line should be obtained; this point will be considered later.

(3) A third possibility has been discussed by Hagen and colleagues (1970). According to this, irradiation does not significantly alter the number of binding sites; but instead, lesions at which RNA synthesis comes to a halt are induced on the DNA. Considerations similar to those of Bresler *et al.* (1964) led to the conclusion that the priming activity of DNA should decrease with dose, following equation (11.1) (Hagen *et al.*, 1969): this means that a plot of A_0/A follows a square-law relationship. The comparison shown in Fig. 86 illustrates that the experimental results of Hagen *et al.* (1970) can be as well described by a linear, as by a quadratic expression. Consequently, no distinction between mechanisms 2 and 3 can be made on the basis of this formal analysis of the dose-response curve.

In order to distinguish which of the various mechanisms contribute significantly to the inactivation of the priming activity the RNA synthesized using irradiated DNA has been studied, as well as the ability of the enzyme to bind to that DNA. As is clearly shown in Fig. 87 a, the length

164

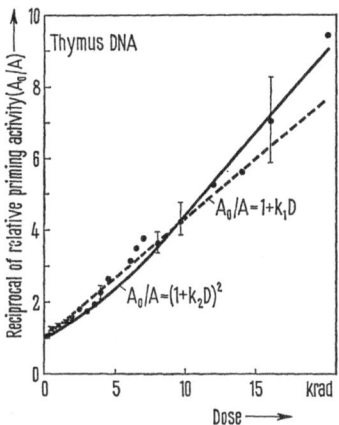

Fig. 86. Inactivation of the priming activity of calf thymus DNA for RNA synthesis by irradiation in aqueous solution (0.5 mg/ml) with ^{60}Co γ-rays. ———— Description of the experimental points by equation (11.1); — — — Description of the experimental points by equation (11.6). (Hagen *et al.*, 1970)

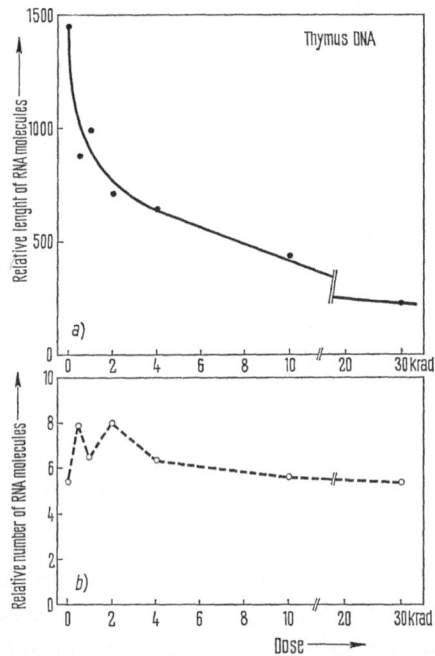

Fig. 87. Length and number of the RNA molecules synthesized on irradiated thymus DNA. a. Relative length of the RNA molecules determined from the ratio of [8-^{14}C]-AMP and [γ-^{32}P]-ATP incorporated. b. Relative number of RNA molecules determined from the incorporation of [γ-^{32}P]-ATP into RNA. (Hagen *et al.*, 1970)

165

of the synthesized RNA chains decreases at approximately the same rate as the total quantity of newly synthesized RNA (see Fig. 85). In contrast, the number of chains is not significantly altered (Fig. 87 b). Radiation damage to the primer DNA therefore alters the RNA synthesis in such a way that only a section from a starting point to the next critical lesion is transcribed, and not the total (cf. mechanism 3). Nevertheless, the other two mechanisms also play a part in the inactivation of priming activity. The sedimentation of RNA in a sucrose gradient showed that after small doses, the number of large RNA molecules decreases relatively more rapidly than that of smaller ones, which supports the first hypothesis. In addition, it has been shown that heavily irradiated DNA has an increased ability to bind RNA polymerase, which supports the second hypothesis (Kröger and Schuchmann, 1966). However, at low doses this latter effect contributes relatively little to the inactivation of the DNA function under discussion.

What, then, is the *nature of the critical lesion?* Inhibition of RNA synthesis can, in principle, be caused by double strand breaks, single strand breaks, base damage or rupture of hydrogen bonds. An indication of the critical lesion is obtained from a comparison of the priming activity of various DNA preparations with their molecular weights (Fig. 88). The DNA was irradiated with γ-rays and UV light, as well as being degraded by DNase and ultrasonics. The unusual form of presentation enables a direct estimate to be made of the effect of double strand breaks on the priming activity since these, in contrast to other lesions, cause a reduction in the molecular weight. Fig. 88 shows no unambiguous correlation between the priming activity of thymus DNA and its molecular weight. Ultrasonic degradation of DNA only interferes significantly with the priming activity at low molecular weights, of the same order as that of the genetic marker. Double strand breaks, which are caused by ultrasonic treatment, therefore cannot be considered as the critical event for the inhibition of priming activity. It is interesting to note that the results of exposure to ultrasonics and DNase lie on the same curve, although the DNase-induced degradation is due to an accumulation of single strand breaks. This also implies that single breaks are not responsible for the inactivation of the priming activity. On the other hand, it is unlikely that a change in the hydrogen bonds would account for the loss of priming activity, since DNA completely denatured by heat still retains about one-half of the priming activity of the native molecule, while a reduction to practically zero can occur after irradiation.

The arguments given so far leave only certain types of base damage as the critical lesions which may lead to the inactivation of the priming activity of DNA after irradiation with UV light. This view is supported by the observation that, at a given molecular weight, the priming activity is inactivated with high efficiency by UV light (Fig. 88), which predominantly induces base damage, but is inefficient for the production of strand

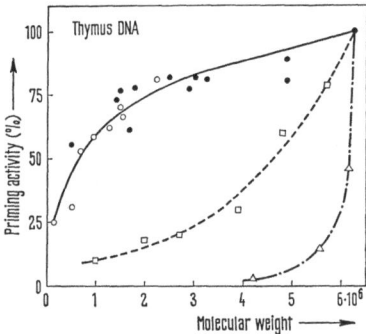

Fig. 88. Priming activity (incorporation of AMP into the RNA) of calf thymus DNA as a function of its molecular weight after exposure to various agents. □ ^{60}Co γ-radiation; △ UV light; ● DNase I; ○ ultrasonics. (Hagen *et al.*, 1970)

breaks. For the action of ionizing radiation the nature of the critical lesion has not been completely resolved. There are some experimental results (compiled by Weiss and Wheeler, 1967), which do not support the view that base changes are the crucial events for the loss of priming activity after γ-irradiation. Single strand breaks formed by DNase attack do not affect the priming activity of DNA to a large extent (cf. Fig. 88); but by this enzymatic degradation one single phosphodiester bond is hydrolized, whereas at the site of a radiation-induced break either the nucleotide base bound to the damaged sugar residue or the whole nucleoside is lost (Simon, 1969). This deficient base in conjunction with a strand break may actually be responsible for stopping the transcription on irradiated DNA (Hagen *et al.*, 1969).

11.5. Enzyme Induction

The test of the priming activity of DNA is not the only method of measuring the inactivation of a transcription process. Enzyme induction is a suitable procedure to be used as an overall test of the process of transcription of a certain marker. If, for example, the synthesis of β-galactosidase is induced in E. coli 15 T$^-$L$^-$, it is found that a linear increase in the amount of enzyme synthesized by the bacteria occurs after a latent interval of about five minutes (Fig. 89). With increasing radiation dose, the slope of the linear portion of the curve (i. e. the quantity of galactosidase synthesized per unit time) decreases. However, only a marginal decrease in the rate of synthesis, relative to the unirradiated control, is observed after low doses (5.5 krad); thus a dose-response curve with a pronounced shoulder is obtained if the rate of synthesis is plotted as a function of radiation dose. The decrease in the amount of enzyme produced probably results from interference with the transcription process; it will be shown in the next

167

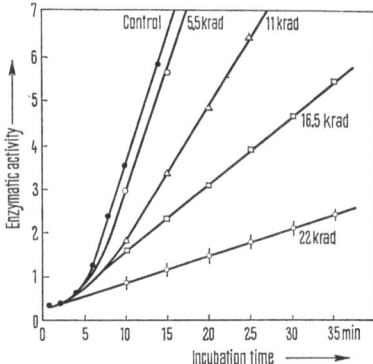

Fig. 89. Inactivation by ^{60}Co γ-radiation of β-galactosidase activity induced in E. coli T^-L^-. The amount of galactosidase per ml of bacterial culture was determined at various time intervals after induction, i. e. as a function of the incubation time in $5 \cdot 10^{-4}$ M thio-β-D-galactopyranoside. (Pollard and Barone, 1966)

section that the later steps in enzyme synthesis, i. e. the actual steps of translation, are much more radiation resistant than enzyme induction. This result is emphasized by the experiments of Pauly (1963), which show that the rates of synthesis of proteins, induced enzymes, and RNA in Bacterium cadaveris decrease exponentially with radiation dose, and the D_{37}-values for these three functions are practically the same (about 30 krad). It may be concluded from this that radiation action probably blocks RNA synthesis, with the inhibition of protein synthesis representing a related secondary effect.

11.6. DNA-mRNA Hybrids

A further property of DNA, also associated with transcription, is its ability to form a hybrid with the corresponding messenger RNA. An example of an investigation of this kind is given in Fig. 90: Robev and Marinova (1967) mixed irradiated DNA from E. coli B with ^{32}P-labelled mRNA also from E. coli B, and after incubation at 78° C for 3¹/₂ hrs, they determined the percentage of hybrid complexes, after alkaline hydrolysis and chromatographic separation of the ribonucleotides. As shown in Fig. 90, only a relatively small effect is observed at low doses, while above about 20 krad the hybridization breaks down completely within a comparatively small dose range. As might be expected, only a relatively small amount of hybridization is obtained if the mixture of two nucleic acids is not incubated, or if an attempt is made to hybridize DNA from B. subtilis with mRNA from E. coli. The form of the dose-response curve suggests that mRNA accepts the DNA as being "complementary" even if it already contains several radiation-induced lesions. A critical point is reached at high doses,

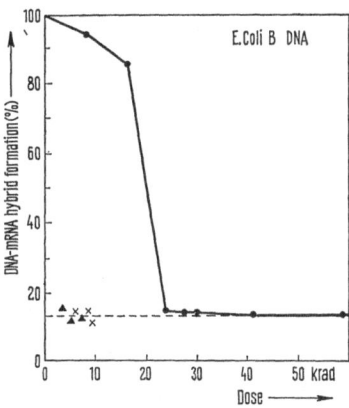

Fig. 90. Inhibition by X-rays of the formation of DNA-mRNA hybrids by irradiation of DNA from E. coli B in aqueous solution (1.2 mg/ml). ● Irradiated DNA of E. coli B hybridized with mRNA of E. coli B after 3.5 hr incubation at 78° C; ▲ DNA from E. coli B hybridized with mRNA of E. coli B without incubation; x DNA of B. subtilis hybridized with mRNA of E. coli B with incubation. (Robev and Marinova, 1967)

due to the accumulation of lesions, beyond which no more hybridization is possible.

11.7. Translation

In this section, an overall survey of the relative radiation sensitivities of the various stages of translation of genetic information from the nucleic acid system to the protein system will be given. This survey will cover the functions of mRNA, transfer RNA (tRNA) and the ribosomes.

a) Messenger RNA

The functions of mRNA can be tested in experiments similar to those carried out on enzyme induction. There is, however, one difference: the galactosidase production is induced first, resulting in the formation of a large number of mRNA molecules, and only then are the induced cells irradiated. In these experiments (Pollard and Barone, 1966), it was found that the enzyme production was not affected by doses of less than 12 krad. This is not surprising since the mRNA molecules, having molecular weights between 10^5 and 10^6 Dalton, represent relatively small targets and are correspondingly radiation resistant; thus, at 12 krad and with a mean hit energy of 60 eV, an inactivation of 0.2 to 2% would be expected. This type of experiment thus clearly indicates that the reduction in enzyme production caused by irradiation of bacteria before induction (cf. Chapter 11.5), cannot be due to a change in the mRNA or in the succeeding stages of translation, but is due to an alteration of the transcription process.

b) Transfer RNA

Transfer RNA is a relatively small RNA molecule, composed of about 70 nucleotides, with a molecular weight of about 25,000. Its function is to bind amino acids, and to initiate their incorporation at the correct site into a protein at the ribosomes. The binding ability of tRNA can be tested in vitro by, for example, incubating yeast tRNA ("soluble RNA") with radioactively-labelled amino acids, precipitating the RNA after several minutes, and determining the quantity of labelled amino acids in the acid-insoluble fraction. If the tRNA is irradiated in the dry state at $-80°$ C with 1 MeV electrons, the binding capacity for various amino acids decreases exponentially with dose (Fawaz-Estrup and Setlow, 1964). However, the slopes of the dose-response curves differ for the different amino acids, and therefore for the various kinds of tRNA. The target molecular weights, determined using the D_{37} and equation (5.5), lie between 6,500 and 23,000 depending on the amino acid (Table 14), and are therefore of the same

Table 14. $37^0/_0$-Doses for the inactivation of the binding ability of transfer-RNA for various amino acids, and the target molecular weight MW_T calculated from them. (Fawaz-Estrup and Setlow, 1964)

Amino acid/tRNA-complex	D_{37} [Mrad]	MW_T
Valine	86	6,500
Methionine	62	9,500
Proline	58	10,000
Isoleucine	46	12,500
Alanine	43	13,500
Leucine	25	23,000

order of magnitude as the true values. The absorption of 50 to 200 eV of radiation energy leads in almost every case to the loss of the ability of the various kinds of tRNA to bind their specific amino acids.

c) Ribosomes

The final reaction step in the translation occurs in the ribonucleoprotein particles, known as ribosomes. In E. coli, the functional particle has a sedimentation coefficient of 70 S and molecular weight of $2.6 \cdot 10^6$ Dalton. Reduction of the magnesium ion concentrations causes ribosomes to split into two subunits, with sedimentation coefficient of 50 S and 30 S. During protein synthesis, tRNA attaches to the 50 S particles and messenger RNA to the 30 S particles. The ribosomal function can be tested as follows: ribosomes extracted from, for example, E. coli are introduced after repeated purification into a suitable in vitro system, containing polyuridylic acid as the mRNA. The corresponding codon, which was the first to be determined

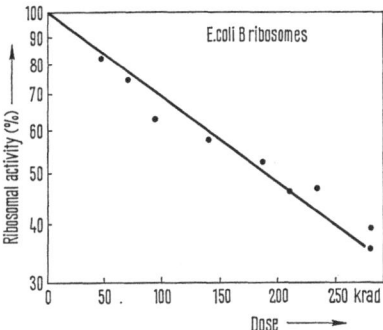

Fig. 91. Inactivation of freeze-dried ribosomes of E. coli B by ^{60}Co γ-radiation; further explanation in the text. (Kućan, 1966)

in investigations of the genetic code, is the sequence UUU, i. e. a sequence of thee uracil residues, which codes for the amino acid phenylalanine. Consequently, if tRNA and phenylalanine are present in the reaction mixture, poly-U initiates the synthesis of polyphenylalanine. If freeze-dried ribosomes are irradiated with ^{60}Co γ-rays then the activity, and therefore the quantity of phenylalanine synthesized per unit time, decreases exponentially with dose (Fig. 91). The D_{37} is 270 krad, from which a target molecular weight of $2.2 \cdot 10^6$ Dalton can be calculated according to equation (5.5). This value is in reasonable agreement with the true molecular weight of the 70 S particle of $2.6 \cdot 10^6$; i. e. ribosomes are also inactivated by a single energy loss event of 60 eV, with a probability of about 1. As both of the ribosomal subunits play an active part in protein synthesis, it is not unreasonable to assume that the inactivation of one subunit will destroy the functional ability of the whole unit.

This concludes the discussion of the most important functions of nucleic acids and their inactivation by radiation. It has been shown that the components participating in translation are considerably more radiation resistant, because of their low molecular weights, than the functions of transcription and replication. Of these latter two functions, RNA synthesis is in turn more radiation-resistant than DNA synthesis, which can be explained by the smaller size of the molecular sections involved in transcription. It must, therefore, be concluded that the critical event is a radiation-induced interference with DNA replication, which is thus the primary cause of reproductive death, e. g. leading to the inactivation of bacteria. The conclusion that DNA is the most important radiation sensitive target in viruses and cells will be supported by numerous other examples given in the following chapters. However, there is also the possibility that transcription may, in certain circumstances, contribute a significant proportion of the radiation damage in cellular systems.

171

This discussion has once again made it abundantly clear that the inactivation event must be investigated by physico-chemical means, and not only by the recording of dose-response curves. This chapter contains numerous examples of different types of dose-response curves, most of which could not be interpreted unambiguously. It may, therefore, have given the impression that dose-response relationships cannot represent the "experiment proper", but only some kind of preliminary investigation which may, in favourable circumstances, lead to the formulation of a working hypothesis. In the past too many scientists have misinterpreted hit and target theories being satisfied with mathematical analyses of their results and, as a consequence, not continuing with the detailed investigation of their conclusions that is necessary in an exact science. This has led to a wide-spread reporting of dose-response curves, the interpretations of which are restricted by the lack of suitable "follow-up" experiments. One of the aims of modern radiation biology is to elucidate the molecular mechanisms of radiation action. This aim cannot, however, be achieved merely by the analysis of dose-response curves, but requires extended investigations using the powerful physico-chemical methods available today.

References

Blok, J.: In: Radiation research. Ed.: G. Silini. Amsterdam: North-Holland Publ. Co. 1967, p. 423.
Bresler, S. E., Kalinin, V. L., Perumov, D. A.: Biopolymers 2, 135 (1964).
— — — Mutation Res. 4, 389 (1967).
Cato, A., Guild, W. R.: J. molec. Biol. 37, 157 (1968).
Fawaz-Estrup, F., Setlow, R. B.: Radiat. Res. 22, 579 (1964).
Fiers, W., Sinsheimer, R. L.: J. molec. Biol. 5, 424 (1962).
Griffith, F.: J. Hyg. 27, 113 (1928).
Guild, W. R., Defilippes, F. M.: Biochim. biophys. Acta (Amst.) 26, 241 (1957).
Guthrie, G. D., Sinsheimer, R. L.: Biochim. biophys. Acta (Amst.) 72, 290 (1963).
Hagen, U., Ullrich, M., Jung, H.: Int. J. Radiat. Biol. 16, 597 (1969).
— — Petersen, E. E., Werner, E., Kröger, H.: Biochim. biophys. Acta (Amst.) 199, 115 (1970).
Haldane, J. B. S.: J. Genet. 8, 299 (1919).
Harrington, H.: Proc. nat. Acad. Sci. (Wash.) 51, 59 (1964).
Hutchinson, F.: In: Biological effects of ionizing radiation at the molecular level. Vienna: Internat. Atomic Energy Agency 1962, p. 15.
Jung, H.: Habilitationsschrift, University of Heidelberg 1968.
— Hagen, U., Ullrich, M., Petersen, E. E.: Z. Naturforsch. 24 b, 1565 (1969).
Kornberg, A.: Enzymatic synthesis of DNA. New York: John Wiley & Sons 1961.
Kröger, H., Schuchmann, L.: Biochem. Z. 346, 191 (1966).
Kućan, Z.: Radiat. Res. 27, 229 (1966).
Lerman, L. S., Tolmach, L. J.: Biochim. biophys. Acta (Amst.) 33, 371 (1959).
Lytle, C. D., Ginoza, W.: Int. J. Radiat. Biol. 14, 553 (1969).
Pauly, H.: Int. J. Radiat. Biol. 6, 221 (1963).
Pollard, E. C., Barone, T. F.: Radiat. Res. Suppl. 6, 124 (1966).
Ravin, A. W.: Advanc. Genet. 10, 61 (1961).
Robev, S., Marinova, Z.: Nature 213, 935 (1967).

Rupert, C. S.: Photochem. Photobiol. **7,** 437 (1968).
— Goodgal, S. H.: Nature **185,** 556 (1960).
Setlow, R. B., Setlow, J. K.: Proc. nat. Acad. Sci. (Wash.) **48,** 1250 (1962).
Simon, M.: Int. J. Radiat. Biol. **16,** 167 (1969).
Sinsheimer, R. L.: J. molec. Biol. **1,** 43 (1959).
Tanooka, H., Hutchinson, F.: Radiat. Res. **24,** 43 (1965).
Watson, J. D.: Molecular biology of the gene. New York: W. A. Benjamin 1965.
Weiss, J. J., Wheeler, C. M.: Biochim. biophys. Acta (Amst.) **145,** 68 (1967).

Chapter 12. The Action of Radiation on Viruses

12.1. Basic Properties of Viruses

Viruses are biological objects which have no metabolism of their own and can replicate only with the aid of the genetic and metabolic systems of a host cell, i. e. they are "parasites at the molecular level". If the host cell is a bacterium they are referred to as bacteriophages, or phages. Since the effects of radiation have been investigated more thoroughly in bacterial viruses than in other systems, a large proportion of the experimental material discussed in this chapter is derived from investigations on bacteriophages.

Although there are numerous books that describe the properties and details of the replication cycles of many viruses (for bacteriophages see, for example, Adams, 1959, and Stent, 1963), a short summary of the main features and the most important terms will be given before considering the response of viruses to the action of radiation. Viruses consist of a protein coat (capsid) of varying complexity, enclosing the carrier of genetic information, i. e. some type of nucleic acid. Most *RNA viruses* contain single-stranded nucleic acid, but a few are known to contain double-stranded RNA. *DNA viruses* have either single or double strands, which may be in the form of a single thread or a ring. Tables 15 and 16 give some representatives of these groups. The process of virus reproduction, known as *infection,* may take various forms depending on the nature of the host cell. It begins with the *penetration* of a host cell by a virus; this problem, which is tackled in a variety of ways by different types of viruses, has been solved in a particularly ingenious manner by T-phages of the intestinal bacterium E. coli. These phages have a "head" containing the nucleic acid and a "tail" that adheres to special "receptors" in the cell wall of the host. Possibly, enzymes contained in the tail digest the cell wall, forming a hole through which the DNA is injected. This is followed by a *latent phase,* which is characterized by the fact that no functional virus can be detected in the host (in E. coli, its duration is about 15 min). In its first stage (known as the *eclipse),* the synthesis of the *early enzymes* occurs. Once these enzymes (predominantly polymerases and nucleases) have been prepared, then *nucleic acid synthesis* commences. In double-stranded viruses this process occurs semi-conservatively, according to the Watson-Crick

174

model, but in RNA viruses the details are not yet clearly defined. The circular single-stranded DNA of the phage $\Phi X\,174$ (which is known to infect protoplasts even without its capsid; see Chapter 11.2) is converted, shortly after penetration, to a circular double strand known as the replicative form (RF), which is also infectious. Several copies of this doublestranded DNA are then produced by semi-conservative replication, and these then act as templates for the synthesis of single-stranded ΦX-DNA (cf. Sinsheimer, 1968).

If a sufficiently large number of newly-synthesized DNA molecules are formed, then synthesis of capsid protein commences, followed by the *maturation of the virus*, i. e. the uniting of DNA with the capsid. The details of the processes of viral assembly are also mostly unknown. The final *release* of viruses from their host cell occurs in many different ways. Most phages decompose the cell wall of the host bacterium enzymatically, causing *lysis* (destruction of the cell), and leading to the release of between 100 and 200 new phages, depending on the type of phage. Other viruses are simply "leaked" out of the cell, or form a bud in the cell wall of the host, which then contracts and separates, leaving the host cell more or less intact.

The T-phages of E. coli. The seven phages of the T-series of E. coli all contain double-stranded DNA, and their response to radiation has been investigated in great detail. They are usually divided into two classes: the even- and the odd-numbered T-phages. Reproduction in the first group, i. e. T2, T4 and T6, is independent of the integrity of the bacterial genome, and there is experimental evidence that this is actually destroyed during infection. Since these phages contain 5-hydroxymethylcytosine instead of the DNA base cytosine, they also differ from the odd T-phages with respect to the eclipse phase, because several enzymes have to be synthesized, e. g. those which convert cytosine to hydroxymethylcytosine (the hydroxymethylases), before DNA synthesis commences. As these steps require additional genetic information, it is not surprising that the evennumbered T-phages contain 3 to 4 times as much DNA as the odd phages (see Table 16). The latter contain cytosine and do not (with the exception of T5) destroy the E. coli genome.

Temperate Phages. Phages which replicate in the manner just described are known as *virulent*. There are, however, cases where the phages lead a "peaceful existence" after penetration of the host, with no virulent multiplication. These are referred to as *temperate phages*, and the host cells as *lysogenic bacteria*. The best known example of this type is the λ-phage; it has a double-stranded circular DNA molecule, and its host is E. coli K12. During the "peaceful" stage (the *prophage* stage) the λ-DNA attaches to the coli genome in the vicinity of the galactose marker "gal" (see Fig. 113), and is then replicated together with the bacterial DNA, and even inherited by the daughter cells after division. The prophages can, however, be "induced" at any time, for example by UV light; i. e. they become virulent and lyse the cells. If the progeny of induced λ-prophages infect new lysogenic coli cells, the "gal"-information of the original host is often found to be transferred, or *"transduced"*, to the new host. λ-phages, because of these and other characteristics, are becoming increasingly favoured as a radiation biological test system. They will be mentioned repeatedly in Chapter 12.4.

Detection of phages. On the basis of the above mentioned details of the infective cycle, phages can be detected as follows: the bacteriophages to be counted are added to an excess of host bacteria, "plated" on nutrient agar plates, and incubated at 37° C for about 15 hours, so that the bacteria form a continuous "lawn". All the bacteria in the vicinity of a site where there was a phage at the time of plating will be lysed by the progeny of that phage; this produces a hole, known as a "plaque", in the layer of bacteria. These plaques, having a diameter of the same order of magnitude as that of bacterial colonies, can be counted with the naked eye. Further details of phage assay may be found in the book of Adams (1959).

12.2. Inactivation of Viruses containing Single-Stranded Nucleic Acids

The viruses containing single-stranded DNA or RNA are the simplest nucleic acid-containing systems, and therefore give a greater possibility of obtaining unambiguous information about the elementary processes of radiation damage from the experimental results. Nevertheless, this section will be relatively short since the action of radiation on single-stranded viruses (with the possible exceptions of phage $\Phi X\,174$ and tobacco mosaic virus) has not as yet been investigated in as much detail as the action on viruses containing double-stranded DNA. A comparison of the *radiation sensitivities* of different types of virus will first be carried out, using the 37%-doses for inactivation (i. e. the loss of plaque-forming ability) obtained under what are essentially "direct action" conditions. However, this is not always possible, since many viruses do not survive desiccation. In these cases, results from experiments carried out in solution in the presence of radical scavengers (such as nutrient broth or histidine) or in frozen suspensions, have to be relied on: the D_{37} values obtained under these conditions are generally under-estimates, since in most cases the sensitivity level of the dry state is not obtained (see Fig. 56). The molecular weights of the targets calculated from the 37%-doses, using equation (5.5), are in relatively good agreement with the molecular weights of the nucleic acids of the corresponding viruses, as shown in Table 15. Certain conclusions can therefore be drawn, but only with some reservations that are necessary with all target analyses of this kind:

1. The whole of the nucleic acid is the radiation sensitive target in viruses containing single-stranded DNA.

2. Under experimental conditions where direct radiation action predominates the sensitivity of the nucleic acid is not measurably affected by the protein coat. This has also been shown by direct measurements, in which the RNA isolated from tobacco mosaic virus, and the DNA from phage ΦX 174, were found to have the same radiation sensitivity as the whole virus (Table 15).

3. The protein coat, which in most viruses represents about a half of the total weight, and therefore absorbs approximately the same amount of

Table 15. Sensitivity to ionizing radiation of viruses with single-stranded nucleic acid. Comparison of the target molecular weight (MW_T) calculated from the 37%-doses (D_{37}) with the true molecular weights of the DNA or RNA. (Ginoza, 1967)

	MW of the DNA and RNA [10^6 Dalton]	D_{37} [krad]	MW_T [10^6 Dalton]	MW_T/MW
DNA-Viruses:				
Phage ΦX 174	1.7	380	1.5	0.9
ΦX 174-DNA	1.7	320—380	1.5—1.8	0.9—1.1
Phage S13	1.7	390	1.5	0.9
RNA-Viruses:				
Tobacco mosaic virus	2.0—2.2	290—300	1.9—2.0	0.9
Tobacco mosaic virus RNA	2.0—2.2	300	1.9	0.9
Phage R17	0.7—1.1	750	0.8	0.7—1.1
Bushy stunt virus	1.5	450	1.3	0.9
Tobacco ring spot virus	1.5	450	1.3	0.9
Tobacco necrosis virus	1.5	670	0.9	0.6
Rous sarcoma virus	10	45—200	2.9—12.9	0.3—1.3
Newcastle disease virus	7.5—33	50	11.6	0.4—1.5

radiation energy as the nucleic acid, does not affect the virus function tested; i. e. in the inactivation caused by direct action, the proportion of viruses with a protein lesion (e. g. those which have lost their ability to attach to the host cell) is unimportant compared with the proportion that contain damaged nucleic acids.

4. In most cases, the absorption of an average amount of energy of 60 eV is sufficient to cause the inactivation of a single-stranded virus.

The credibility of these "statements" is basically, of course, only as great as that of the target theory itself. Nevertheless, with the due care that is required in discussions of target analysis, it can be concluded that the different radiation sensitivities of the various single-stranded viruses reflect primarily their different nucleic acid contents rather than a basic difference in their sensitivities; i. e. the larger the DNA content, the more sensitive the virus. The discussion of the target in enzymes (see Fig. 28), led to a similar conclusion.

The inactivation of a virus means, of course, its inability to complete all the complex reaction steps involved in a normal replication cycle (cf. Chapter 12.1). The question of the reaction steps to which an inactivated phage can proceed has been investigated by Lytle and Ginoza (1969a) using phage ΦX 174. After being irradiated in frozen nutrient broth, attachment to the host bacterium was not measurably affected. About 70% of the inactivated phage were still able to inject their DNA into their host. This percentage was comparable with the fraction of phages that had no strand breaks in their DNA (Lytle and Ginoza, 1969b; cf. Chapter 11.2). Of the damaged single strands injected, more than 80% were converted at least into partial RF-DNA. The damaged DNA could not prevent the transcription or replication of undamaged DNA by competition, nor could it kill the host bacterium. These observations probably result from the fact that the DNA of inactivated phages cannot become attached to the special sites on the host cell membrane where RF-replication takes place (cf. Sinsheimer, 1968). Consequently, it cannot kill the bacterium by preventing host cell synthesis which presumably occurs at the same replication sites. None of the phage-specific functions, as measured by complementation, were expressed by inactivated phages although RF-like structures were made. Thus the primary cause of γ-ray inactivation appears to be the physico-chemical damage affecting developmental steps prior to the expression of phage-specified functions, probably by preventing the attachment of RF-DNA to the replication sites on the membrane of the host cell.

The only information available which might help to answer the question as to the *physico-chemical nature of the lethal event*, is that presented in the discussion of radiation action on the infectious ΦX-DNA (Chapter 11.2). In addition to breaks in the single-stranded DNA, which are lethal in every case, base changes would also be expected to contribute to the inactivation. The magnitude of this contribution is approximately three-quarters of the total effect (Lytle and Ginoza, 1969b); the inactivation probability for base changes is not yet known accurately.

12.3. Inactivation of Viruses containing Double-Stranded DNA

Most viruses containing double-stranded nucleic acids have a DNA genome. Only three types of virus containing double-stranded RNA are known (see Kaplan, 1968), and as their responses to ionizing radiation have not yet been adequately examined, this discussion has to be restricted to DNA viruses. The discussion will again commence with a comparison of the DNA molecular weight with the MW_T-values calculated according to equation (5.5), (see Table 16). The ratio MW_T/MW is known as the inactivation probability, and is also commonly referred to as the killing efficiency; it lies between 0.05 and 0.1, except for the first three types of viruses which will be discussed separately. On the basis of their DNA content, the double-stranded viruses are considerably less sensitive to radiation than those containing single-stranded nucleic acids, the killing-efficiency of which was shown to be about 1. Three possible explanations of this higher resistance can be given:

1. The physico-chemical nature of the inactivating event in a double strand virus differs from that in a single strand virus, and occurs less frequently.

2. Repair mechanisms reduce the radiation sensitivity of these double-stranded viruses.

3. The DNA molecules, having molecular weights 1 to 2 orders of magnitude greater than those of single-stranded viruses, have a sensitive section ("critical target") in which damage leads to inactivation, while absorption of radiation energy in other parts of the DNA is of no consequence.

These hypotheses are based on the assumption that the DNA is the only sensitive target within the viruses. On the basis of all that has been said in this chapter and will be said in Chapter 14, there is little doubt about the correctness of this assumption especially for direct radiation action. It appears unlikely that the third hypothesis is valid, although certain experimental results do seem to support it (see Ginoza, 1967). The discussion of the other two hypotheses, which will now be carried out in detail, begins with the question as to the nature of the inactivating event.

a) Single Strand Break as the Inactivating Event

The high *radiation sensitivity of phage* α (see Table 16) calls for a closer examination. The double-stranded nature of this Bacillus megaterium-phage can be considered as having been established (Aurisicchio et al., 1962). The

179

Table 16. *Sensitivity to ionizing radiation of viruses with double-stranded DNA. Comparison of the target molecular weight (MW$_T$) calculated from the 37%-doses (D$_{37}$) with the true molecular weight of DNA.* (Ginoza, 1967)

	MW [10^6 Dalton]	D$_{37}$ [krad]	MW$_T$ [10^6 Dalton]	MW$_T$/MW
Phage α	30	22—27	22—26	0.7—0.9
Polyoma virus	3	500	1.2	0.40
RF-DNA of ΦX 174	3.4	780	0.7	0.21
Phage T1	30	320—570	1.0—1.8	0.03—0.06
Phage T2	130	55—100	5.8—10.5	0.04—0.08
Phage T4	130	100	5.8	0.04
Phage T7	42	150	3.9	0.09
Phage λ	31	380	1.5	0.05
Phage 22	39	140	4.1	0.11
Phage BM	25	210	2.8	0.11
Adeno virus type V	66	77	7.5	0.11
Shope papilloma virus	14	480	1.2	0.09
Vaccinia virus	156	80	7.2	0.05

37%-dose for its inactivation in nutrient broth or histidine solution lies between 22 and 27 krad (Celano et al., 1960; Freifelder, 1966), which gives, according to equation (5.5), a MW$_T$-value of 22 to 26 millions, and thus since the true DNA molecular weight is 30 million, gives a killing efficiency of 0.7 to 0.9. This value is in agreement with the result of Aurisicchio et al. (1962), who estimated a killing efficiency of 0.5 to 1.0 per disintegration for the inactivation of phage α by ^{32}P-decay.

Freifelder (1966) was able to show by ultracentrifugation that the proportion of DNA molecules with single strand breaks increases with dose in approximately the same manner as the inactivation of α-phages. A more detailed analysis of the results showed that a break in one of the two strands need not necessarily lead to inactivation. It has to be assumed either that one particular strand of the two must be broken, or alternatively, that in phage multiplication within the cell, DNA synthesis commences with a DNA strand chosen at random from the two possible strands; if this strand contains a break, then no new phages can be produced.

The two possibilities cannot be distinguished on the basis of Freifelder's results, but one of the two DNA strands of α-phage was shown to be heavier than the other, due to its special base composition. After denaturation of the DNA, the two strands can be separated by centrifugation in a CsCl density gradient. Radioactively labelled mRNA, which is synthesized 15 to 20 minutes after the infection of B. megaterium with α-phages, forms hybrids only with the heavy strand of the denatured α-DNA, and has a base composition analogous to that of the light strand. This is exactly the result that would be expected if the heavy strand alone serves as the matrix for DNA synthesis (Tocchini-Valentini et al., 1963). Accordingly, it is reasonable to assume that the inactivation of extracellularly irradiated α-phages occurs when the heavier of the two DNA strands is broken.

The results described here are by no means typical for viruses containing double-stranded DNA, but represent an exception, since phage α could be considered as a "disguised" single strand phage. It is possible that the *polyoma virus* with its relatively high inactivation probability of 0.4 is also an exception, especially since denaturation of the double-stranded circular polyoma-DNA does not necessarily lead to a loss of infectivity, but may even enhance it significantly (Weil, 1963).

Another exception, as far as its relative radiation sensitivity is concerned, is the *replicative form of ΦX 174-DNA* (see Table 16). This RF-DNA has twice the molecular weight, but only half the radiation sensitivity of the single-stranded DNA (Ginoza and Miller, 1965), resulting in a killing-efficiency of 0.21. Similar results have been obtained using UV light (Yarus and Sinsheimer, 1964) and ^{32}P decay (Denhardt and Sinsheimer, 1965). Taylor and Ginoza (1967), using zonal centrifugation, found a ratio of single to double strand breaks of 38:1 in irradiated RF-DNA. This means that only a small proportion of the inactivation of RF-DNA can be attributed to double strand breaks, and it is probably due predominantly to single strand breaks and base damage.

b) Double Strand Break as the Inactivating Event

An indication of the type of event producing inactivation in most of the double-stranded viruses listed in Table 16, is obtained from the following comparison: a killing-efficiency of 0.05 to 0.1 per 60 eV corresponds to a G-value of 0.08 to 0.17. This value is in agreement with the results of Chapter 10.4, where it was shown that in the irradiation of DNA, either under essentially direct action conditions (i. e. in the dry state, in cells or as a nucleo-protein gel) double strand breaks are produced with yields of $G = 0.1$ to 0.15. On the basis of this agreement, it seemed worth while to examine the correlation of the inactivation of double-stranded viruses with breakage of the two DNA strands.

Phage T7 in phosphate buffer give a survival curve with an initial shoulder (Fig. 92). In contrast, the inactivation in nutrient broth or in 10^{-3} M histidine solutions gives purely exponential dose-response curves with identical slopes ($D_{37} = 84$ krad). The addition of cysteine to the histidine solution results in a further reduction in radiation sensitivity (see also Fig. 47). The D_{37} correspondingly increases to 175 krad (Fig. 92). If the fraction of DNA molecules remaining unchanged after irradiation in buffer is determined using an analytical ultracentrifuge, a 1:1-correlation between this fraction and the percentage of surviving phages is found (Fig. 93). A linear extrapolation of the results obtained in histidine or nutrient

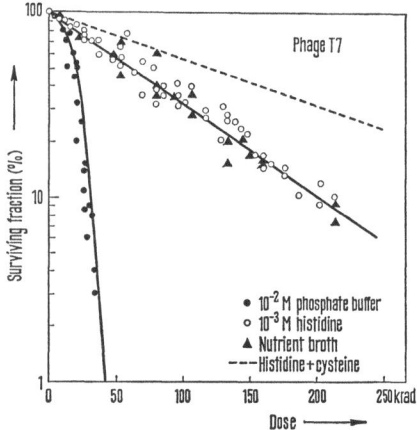

Fig. 92. Inactivation of T7-bacteriophages by X-rays in phosphate buffer (pH 7.8), 10^{-3} M histidine solution, and nutrient broth, as well as in anaerobic histidine solution with the addition of cysteine. (Freifelder, 1965)

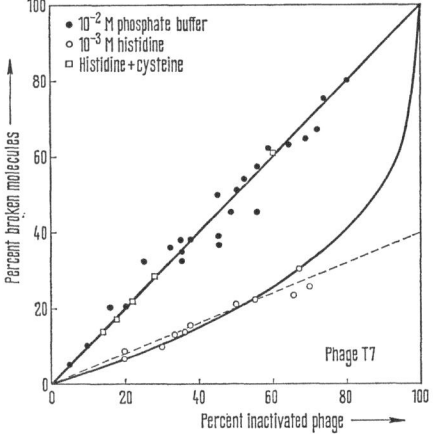

Fig. 93. Comparison of the fraction of broken DNA molecules with the percentage of inactivated phage T7 exposed to X-rays in phosphate buffer (pH 7.8), 10^{-3} M histidine solution, and anaerobic histidine solution with the addition of cysteine. (Freifelder, 1965)

broth gives an estimate that about 40% of the T7-phage inactivation is caused by double strand breaks (Fig. 93, broken line). It would be more correct to plot the proportion of unbroken molecules on a semi-log scale, against the dose; a straight line is then obtained with a D_{37} of 270 krad, from which it is calculated that 31% of the inactivation of T7-phages is caused by double strand breaks (Fig. 93, bent line). On the basis of similar measurements on T1-phages, about 20% of lethal events were attributed to double strand breaks (Bohne et al., 1970). After irradiation in a histidine solution containing cysteine, a 1:1-correlation between the number of DNA molecules containing double strand breaks and the number of inactivated T7-phages is once again obtained (Fig. 93). Since protection by a factor of 2 is observed relative to pure histídine solutions under the same experimental conditions (Fig. 92), the "other lesions", which in nutrient broth or histidine solutions cause 60 to 70% of the observed inactivation, do not occur in the presence of cysteine; in addition there is a simultaneous small increase in the absolute number of double strand breaks.

The shoulder on the dose-response curve obtained by irradiation in buffer solution suggests that the attack of water radicals predominantly causes single strand breaks, and that only when these accumulate with increasing dose does the probability for double strand breaks increase. This mechanism for the production of double strand breaks in dilute aqueous solutions has already been discussed in detail in Chapter 10.4 (see Fig. 74). There is, however, a discrepancy, since Fig. 92 suggests that a double strand break occurs if two single breaks are separated by about 100 base pairs, but this mechanism seems rather improbable. In addition to the accumulation of single strand breaks, an inactivation mechanism similar to that observed by Dewey and Stein (1968), after the exposure of phage T7 in aqueous solution to hydrogen atoms, may be involved. Under these conditions, atomic hydrogen seems to cause a release of the DNA from the phages into the medium, where it is rapidly degraded by further radical attacks. The kinetics of this type of interaction are exponential, as was also found in the exposure of dried T1-phages to H atoms (Jung and Kürzinger, 1968).

Single breaks are relatively ineffective for the inactivation of these systems, as was shown by the investigation of typical double-stranded viruses. This has also been confirmed by the incorporation of radio-phosphorus into phage DNA; ^{32}S is formed when the ^{32}P decays, and it can be assumed that the resultant sulphur-ester bond is hydrolyzed, leading to a single strand break. It has been shown for numerous types of phages that a ^{32}P disintegration leads to inactivation with a probability of about 0.1 (Stent and Fuerst, 1960). This low killing efficiency has been interpreted as indicating that single strand breaks do not cause inactivation of phages containing double-stranded DNA, while in about 1 in 10 of the decays, the recoiling ^{32}S nucleus also breaks the complementary DNA strand. This apparently reasonable hypothesis for phage inactivation has not been confirmed unambiguously by direct measurement. According to Ikenaga (1968) the percentage of inactivated T1-phages correlates with the proportion of DNA molecules containing a double strand break, but according to Reslova and Drobnik (1968) the number of double breaks is at least one order of magnitude smaller than the number of phages inactivated by ^{32}P decay.

Another mechanism which contributes to the inactivation of irradiated phages is the formation of cross-links within a DNA molecule. Such *intramolecular cross-links* can be detected by sedimentation analyses similar to that shown in Fig. 76 for the case of intermolecular cross-linking. However, this process contributes only about 3% of the inactivation of T1-phages after irradiation in nutrient broth. Under these conditions, a comparable proportion (about 3%) of the inactivated T1-phages contain intermolecular cross-links between their DNA and their protein coat (Bohne et al., 1970). In conditions where direct radiation action predominates, *attachment* of inactivated T2-phages to their host bacteria is not measurably affected (Watson, 1950), whereas after irradiation in buffer up to 80% of the inactivated phages were unable to adsorb on their host E. coli (Watson, 1952).

In summary, it can be concluded that double strand breaks in the DNA of bacteriophages lead in every case to the inactivation of the phage. This is the same conclusion that was arrived at in Chapter 5.3, from calculations based on the track segment method (see Fig. 33), according to which the inactivation of phage T1 occurs when an energy loss event consisting of at least two ionizations occurs within a distance of about 12 Å. This was interpreted by identifying the distance of 12 Å with the average diameter of the DNA helix (16 to 20 Å), and by assuming that one ionization occurs in each strand, i. e. a double strand break was postulated.

It was shown that under certain experimental conditions (e. g. in histidine), not all of the observed inactivation can be attributed to double strand breaks; however, the nature of the other inactivating lesions is not yet completely understood. It has nevertheless been established that phages containing double-stranded DNA can, with certain exceptions, withstand 10 to 20 sublethal lesions (single strand breaks and base changes) without being inactivated, while in a single-stranded virus the same lesion is lethal in almost every case. One possible explanation for these observations is that sublethal lesions in double-stranded DNA do not interfere with the replication process, and are simply "passed by". Another possibility is that they are repaired by the host cells, and this point will now be considered in more detail.

12.4. Repair of Radiation Damage in Viral DNA

The most important repair processes known are predominantly enzymatic, and are under the genetic control either of the virus itself, or of the host cell. However, most of the repair processes, the mechanisms of which are known in some detail, can eliminate only *UV lesions*. It therefore seems advisable to examine these processes first of all, before concentrating on the problem of repair of damage caused by ionizing radiation. In the context of repair, the term *reactivation* is commonly used. While repair refers to the actual molecular process of the elimination of damage, reactivation refers to the consequence of such repair, such as an increase in the survival rate. However, this convention is not always respected in the literature.

a) Photoreactivation

Photoreactivation refers to the elimination of UV damage in DNA by post-irradiation "exposure" to light of wavelengths 300 to 400 nm. It is an enzymatic process, as will be shown in the detailed discussion of Chapter 13.4, and the enzyme responsible is present in many cells and organisms. Since the damaged DNA must already be bound to the enzyme during the reactivating exposure, photoreactivation in viruses is observed only after intracellular exposure. It occurs via the splitting of a thymine dimer, and successful repair occurs in viruses with double-stranded DNA (Fig. 95) as well as in single-stranded ΦX-DNA (Winkler, 1964a). Although the enzyme, which can be isolated from yeast and coli bacteria, does not reactivate irradiated RNA (which is reflected in the fact that RNA phages of E. coli cannot be reactivated), photoreactivation is observed in various RNA-containing plant viruses when they are exposed in plant leaves. It is, however, possible that this is due to another enzyme, or that a process known as indirect photoreactivation occurs (see Rupert and Harm, 1966).

b) Host Cell Reactivation

Host cell reactivation (hcr) is also an enzymatic process, which is directed by the host cell but occurs without the action of light. It is very similar to dark reactivation of bacterial DNA, which is discussed in Chapter 13.4. The coli phages λ, T1, T3 and T7, and many phages of other bacteria, can be reactivated. No host cell reactivation occurs with the autonomous phages T2, T4, T5 and T6, which destroy the genome of the host cell after infection. Another system which cannot be reactivated is the single-stranded ΦX-DNA, and this again demonstrates the similarity between host cell reactivation and dark repair in bacteria, which is only possible when the corresponding complementary strand is still intact. Fig. 94 shows an example of host cell reactivation for UV-irradiated T1-phages. If the inactivation of phages is measured by plating on reactivating strains of E. coli (K12S or B), then the dose-response curves have much lower slopes than those obtained with strains that cannot reactivate (E. coli K12S hcr⁻ or B$_{s-1}$). The fact that the slope of the hcr⁺ curve is larger initially than it is at high doses, indicates the occurrence of two different radiation sensitivities. This can be explained by the fact that only a certain proportion of the host bacteria are in a suitable metabolic state to carry out reactivation. Extrapolation of the hcr⁺ line from high doses to $D=0$ (Fig. 94), shows that this fraction is of the order of 30% in an asynchronous bacterial culture. If, however, the DNA synthesis of the host bacteria is inhibited

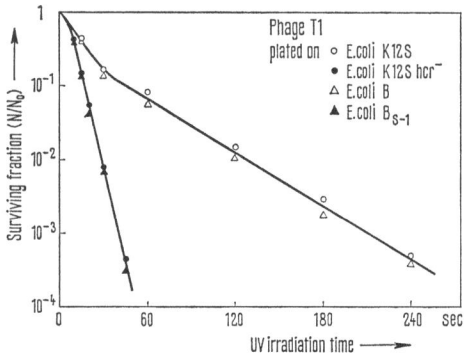

Fig. 94. Inactivation curves of UV-irradiated T1-bacteriophages after plating on host bacteria with different capacities for host cell reactivation. (Harm, 1963 a)

by depletion of the nutrient medium, practically all the hcr+ bacteria carry out host cell reactivation (Sauerbier, 1962).

The fact that the inactivation curves in Fig. 94 are practically exponential, whilst having different slopes, implies that the number of lesions repaired is proportional to the number of lesions produced. It will be shown later that a similar response is observed in the exposure to ionizing radiation of E. coli mutants with differing sensitivities (see Fig. 101). In contrast, capacity for repair of UV damage in bacteria seems to have no influence on the final slope of the inactivation curve, but manifests itself instead by a pronounced shoulder on the dose-response curve (Fig. 102). It can, therefore, be concluded that there is a maximum number of reparable UV lesions in bacteria (see Chapter 13.2). This seems to imply that the dark repair in bacteria is not similar in all respects to host cell reactivation of viral DNA, which may be a "simplified" mode of dark repair.

Although neither the ΦX-phage nor its infectious single-stranded DNA, can be reactivated, both double-stranded RF-DNA (Yarus and Sinsheimer, 1964) and intracellularly irradiated ΦX-phages (Sauerbier, 1964 a) undergo host cell reactivation. Rörsch and colleagues (1964) showed that hcr also occurs in vitro. They incubated UV-irradiated RF-DNA of phage ΦX 174 with a cell-free extract of M. lysodeikticus, and subsequently infected hcr⁻ protoplasts of E. coli. An increase in the survival rate relative to untreated DNA was obtained, although it was not as large as that obtained by the infection of hcr+ protoplasts. Thus reactivation in vitro seems to be slightly less effective than that in hcr+ cells.

c) V-Gene Reactivation

Autonomous coli phages show no host cell reactivation (see above), but it is known that one of them (T4) has an enzymatic repair capacity regulated

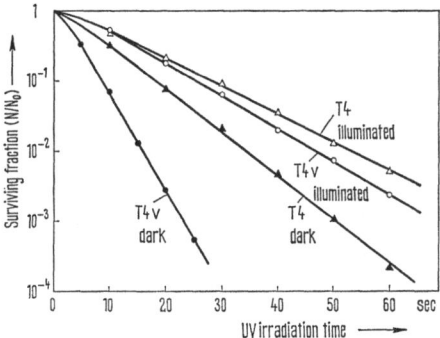

Fig. 95. Inactivation curves of UV-irradiated bacteriophage T4 and T4v with and without subsequent photoreactivation. (Harm, 1963 a)

by the phage genome: v-gene reactivation. The UV sensitivity of mutant T4-phages carrying a v-gene defect (T4v) is twice that of the wild type (Fig. 95). In addition, Fig. 95 shows that photoreactivation and v-gene reactivation are independent processes. However, their action is only partially additive, for the increase in resistance due to photoreactivation is smaller in the wild type than in T4v mutants, since in the wild type a large proportion of the UV damage that is capable of being repaired by photoreactivation has already been repaired by v-gene reactivation. It should be noted that the T4v mutant has approximately the same radiation sensitivity as the T2 wild type, and that the v-gene can be transferred to the T2-phages by crossing.

d) X-Gene Reactivation

The UV sensitivity of the phage T4 is not controlled only by the v-gene. A mutation in the x-gene, which is in the functional position in the wild type, also causes an increase in radiation sensitivity. T4x has a sensitivity lying between those of the wild type and T4v. The action of the v- and x-genes on the UV sensitivity is additive. An interesting observation is that the x-mutation affects genetic recombination: the rate of recombination in the cross-infection of phages (infection of a bacterium by two or more different phage mutants) decreases if all the phages carry the x-mutation (Harm, 1964). The fact that the T4x mutants are also more radiation sensitive indicates that recombination may play a certain part in the repair of radiation lesions; more will be said about this in the discussion of bacteria (Chapter 13.6). In this context, it should be noted that the temperate phage λ, as well as the autonomous phage T4, carries recombination genes which have some influence on the radiation sensitivity. Some experiments on these interesting phenomena are described in the excellent review article by Haynes et al. (1968).

e) Multiplicity Reactivation

According to the definition of inactivation, a phage inactivated by radiation has lost its ability to form intact progeny, but it may nevertheless be capable of carrying out certain important biological functions, such as adsorption to a host bacterium, injection of DNA, and some of the initial processes of infection which, in the case of the autonomous phages, may even lead to the destruction of the host genome. It is not surprising, therefore, that several irradiated phages infecting the same cell can cooperate in a manner leading to the production of functional progeny. This phenomenon is known as *multiplicity reactivation*, and has been examined in numerous experiments and extends to all the T-phages and λ-phage as well as phages of other bacteria. Multiplicity reactivation is observed after treating the phages with UV light or ionizing radiation, if the phage DNA reaches the interior of the bacterial cell. This shows that this type of reactivation is not restricted to specific DNA lesions, but is based on the exchange of genetic material by recombination. If complexes of E. coli bacteria and T4-phages are irradiated with a constant UV dose at different times after infection, then when a bacterium is infected with just one phage, the survival rate remains constant for 4 to 5 minutes and increases only when DNA synthesis commences (Fig. 96, Curve A). Multiple infection leads

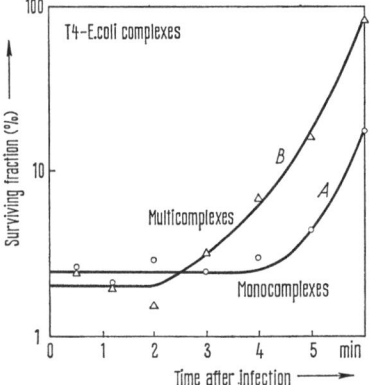

Fig. 96. Surviving fraction of T4-bacteriophages after intracellular irradiation in E. coli bacteria at different times after infection with a constant UV dose. Curve A: Infection of the host bacteria with only one phage each. Curve B: Infection of the host bacteria with two phages on average. (Symonds, 1962)

to a noticeable increase in the survival rate after only two minutes, i. e. in the eclipse phase, as a result of multiplicity reactivation (Fig. 96, Curve B). In a similar reactivation mechanism, known as *cross reactivation* (marker rescue), a host cell is infected by both unirradiated and UV-irradiated phages. However, the term reactivation hardly seems to be justified in this case, since this reactivation effect is a transfer of genetic material from the irradiated phages into the genome of the accompanying intact phages by recombination, and not a true reactivation of the irradiated phages. More detailed information can be obtained from the article of Rupert and Harm (1966).

f) UV Reactivation

As well as the specific reactivation processes considered so far, there are a number of other effects which can cause reversion of the lethal damage in viral DNA. The mechanism of what is known as UV reactivation will be considered in this context. This term covers two different repair processes which, however, have no causal relationship. In Chapter 3.4 (Fig. 14), the continuously decreasing slope of the UV inactivation curve of phage T7 with increasing dose, was attributed to the elimination of radiation-induced lesions by the absorption of a second UV quantum. As was shown in Chapter 10.4, this is caused by the splitting of radiation-induced dimers by the radiation itself. The same mechanism is also responsible for the increase in the transformation rate in Haemophilus influenzae, caused by a second irradiation at 240 nm (see Fig. 84).

Another phenomenon, which is also occasionally referred to as UV reactivation, is made apparent by the following observation: if the host

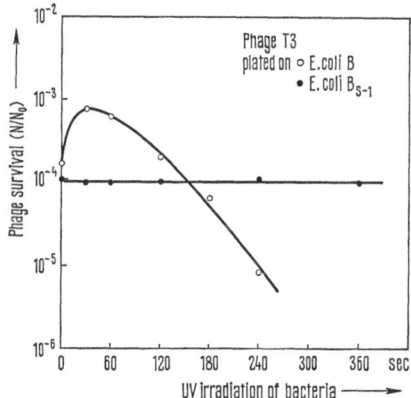

Fig. 97. Surviving fraction of T3-bacteriophages after irradiation with a constant UV dose and plating on host bacteria which had been UV-irradiated prior to infection for different lengths of time. ○ Host: E. coli B; UV irradiation of the T3-phages: 210 sec. ● Host: E. coli B_{s-1}; UV irradiation of the T3-phages: 60 sec. (Harm, 1963 b)

cells are irradiated with increasing doses of UV light before infection with UV-inactivated T3-bacteriophages, then an increase in the T3 survival rate is found in hcr⁺ cells after low doses (Fig. 97). In hcr⁻ bacteria however,

189

the pre-irradiation of the host cells does not influence the number of surviving phages (Harm, 1963 b). This curious result can be explained by the
assumption that a small UV dose induces a physiological state in the host
cell which enables effective host cell reactivation to take place.

g) The Repair of Lesions Induced by Ionizing Radiation

The maximum rate of reactivation detected in bacteriophages after exposure to ionizing radiation was found to amount to 20% of the total
induced lethal damage (Sauerbier, 1964 b; Winkler, 1964 b). This seems to
indicate that none of the mechanisms considered so far is capable of repair-
ing typical lesions induced by ionizing radiation (e. g. single strand breaks)
to such an extent that the relatively high radiation resistance of the double-
stranded viruses (relative to their DNA content) could be explained. This
leaves only the conclusion that the repair of single strand breaks is an
obligatory process during the processes of replication of phage DNA. In
this situation, the repair of a single break in a virus would be linked with
the basic process of reproduction. In fact, numerous observations over the
past few years provide evidence of the fundamental importance of strand-
linking enzymes, such as polynucleotide-ligase (see the review article by
Howard-Flanders, 1968). For example, Weiss and Richardson (1967) succeeded in isolating, from T4-infected coli bacteria, a ligase which could be
used to repair single strand breaks produced in T7-DNA by DNase. Boyce
and Tepper (1968) provided direct evidence of repair of radiation-induced
single strand breaks in the covalent circular DNA of the temperate coli
phage λ; the observed repair is independent of the radiation sensitivity of
the E. coli strain used. This confirms the supposition that the repair of
viral DNA occurs via a reaction which regulates a vital metabolic process,
i. e. the formation of 3',5'-phosphodiester bonds of DNA. In contrast, a
differentiated repair mechanism for single strand breaks in cellular DNA
is encountered in bacteria (Chapter 14.5). Taken as a whole, the repair
processes occurring on viral DNA seem to be simpler than those occuring
on the DNA of the host bacteria, but no less effective.

12.5. BU Effect

Specific chemical changes in the DNA may also influence the radiation
sensitivity. The most significant modification of this kind known at the
present time is the incorporation of halogenated base analogues into DNA.
Representatives of this class of compounds include chlorouracil, bromouracil
(BU) and iodouracil, which differ from thymine only by the substitution
of halogen atoms for the methyl group, and are incorporated into DNA
in place of thymine. The incorporation of these base analogues into DNA
always results in sensitization, for ionizing radiation and UV light as well
as for ^{32}P decay; in the latter case, even if the analogue is in the DNA

strand which does not contain radiophosphorus (for more information, see Kaplan, 1968). The fluorouracil analogue is, in contrast, incorporated into RNA in the place of uracil and not into DNA, and it therefore causes sensitization only in RNA-containing viruses (Becarevic et al., 1963).

To incorporate the most frequently used analogue, bromouracil, into phage DNA, the thymine synthesis of the host bacterium, grown in a minimal medium, is blocked (e. g. by aminopterin) and the thymine-depleted cells are subsequently transferred to and grown in a BU-containing medium. After a few generations, the bacteria are infected by phages, which necessarily results in incorporation of BU into the DNA of the phage progeny.

a) UV Light

Part of the sensitizing action of BU can be explained by the particular physico-chemical properties of the BU-DNA. Setlow and Boyce (1963) are of the opinion that the increased sensitivity of BU-substituted T4-phages at higher UV wavelengths (approximately 300 nm) can be attributed almost completely to the higher optical absorption of the bromodesoxyuridine compared with thymidine. However, another explanation of the sensitizing action of BU must be found for wavelengths below 280 nm. Numerous experimental observations indicate that in BU-DNA the repair of UV lesions is strongly inhibited, possibly by the formation of other non-reparable photoproducts. This applies to the v-gene reactivation; as Stahl et al. (1961) have shown, the sensitivity after BU-substitution depends on the functioning of this gene. On the other hand, the BU-substituted T1-phage plated on the hcr⁻ strain E. coli B_s has approximately the same sensitivity as the unsubstituted phage (Fig. 98), while when the wild strain E. coli B (hcr⁺) is used, the normal phage is considerably more resistant to radiation than the BU-T1 phage. It is worth mentioning that the different absorption by BU is not responsible for the effect shown in Fig. 98 (wavelength of the radiation: 254 nm). The photoreactivation process is also strongly inhibited in BU-containing DNA, as was shown by Stahl et al. (1961) for T2 and by Hotz (1963) for T1-phages. Furthermore, it is observed that after UV irradiation in the presence of cysteamine, followed by plating on E. coli B_{s-1}, both the survival rate and the photoreactivation of BU-substituted T1-phages reach approximately the same levels as those of the thymine-phages (Hotz, 1964). Since the UV sensitivity of normal T1-phages is practically unaffected by cysteamine under these conditions, the BU lesion must be different from the normal photoproducts. According to Hotz, cysteamine may prevent the extension of the BU damage to adjacent regions, in particular to the complementary strand. Secondary radical reactions of the UV-irradiated BU-DNA probably lead to the destruction of deoxyribose, and therefore to the formation of single and double strand breaks (Hotz and Reuschl, 1967; Hotz, 1970).

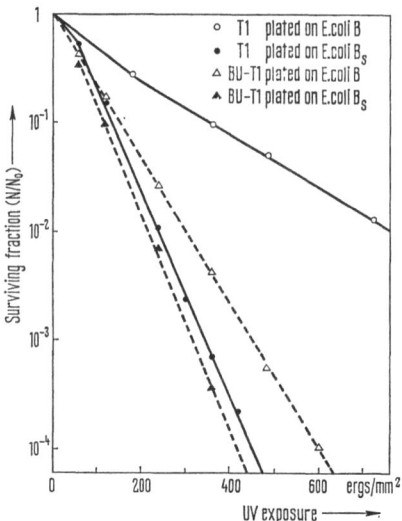

Fig. 98. UV inactivation of T1-bacteriophages with normal and bromouracil-substituted DNA (BU-T1) after plating on host bacteria with different capacities for host cell reactivation. (Howard-Flanders et al., 1962)

b) Ionizing Radiation

An explanation of the increased sensitivity of BU-substituted DNA to ionizing radiation could be based on the high radiation-chemical lability of bromouracil. Müller et al. (1963) found, by means of quantitative ESR-measurements, that the production of free radicals requires much less energy in bromouracil than in thymine; this does not, however, allow any conclusions to be drawn as to the radiation biological significance of BU lesions. ESR-studies by Hüttermann and Müller (1969) on irradiated single crystals of 5-halogen-uracil compounds have shown that of the identifiable types of lesion, the COH radical may be particularly important; it is formed by the addition of hydrogen to the $C_{(4)}$ carbonyl group. In this reaction the hydrogen bond normally occurring at this point in the DNA is destroyed, and this damage may develop into something like a localized denatured zone, leading to labilization of the DNA structure.

The direct action of radiation seems to be responsible for the BU effect, since neither in the irradiation of aqueous solutions (Tanooka, 1964; Freifelder and Freifelder, 1966), nor in exposure to atomic hydrogen (Jung and Kürzinger, 1968) does the incorporation of BU into bacteriophages cause sensitization. This result supports the hypothesis that the "direct" alteration of the position $C_{(4)}$ is responsible for the additional damage in BU-DNA, since the COH radical occurring after γ-irradiation is not observed after exposure to atomic hydrogen, either in halogenated base analogues or any of the other DNA components (Herak and Gordy, 1966).

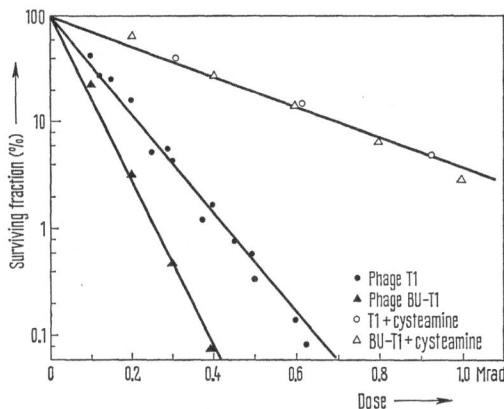

Fig. 99. Inactivation of T1-bacteriophages with normal and bromouracil-substituted DNA (BU-T1) by ^{60}Co γ-radiation in nutrient broth and in the presence of cysteamine at 0.1 M concentration. (Hotz and Zimmer, 1963)

An interesting aspect of the BU effect is emphasized by the experiments of Hotz and Zimmer (1963), which show that in the irradiation of dry T1-phages the presence of cysteamine does not eliminate BU sensitization, but only the "normal" cysteamine protection is observed with BU-phages. In contrast, with γ-irradiation in nutrient broth, the cysteamine compensates completely for the bromouracil effect; i. e. in this system BU-T1 and T1 are both equally sensitive (Fig. 99). This compensation for, or prevention of, the BU effect thus seems to be due to a chemical reaction which can only occur in a liquid medium.

It is unfortunate that at present there are no further observations that can lead to a better understanding of the bromouracil effect in viruses. A further investigation of this effect is without doubt necessary, since it should yield a number of interesting results leading to a better understanding of the molecular processes involved in phage inactivation.

References

Adams, M. H.: Bacteriophages. New York: Interscience Publ. 1959.

Aurisicchio, S., Coppo, A., Frontali, C., Graziosi, F., Toschi, G.: Nuovo Cimento 25, 35 (1962).

Becarevic, A., Djordjevic, B., Sutic, D.: Nature 198, 612 (1963).

Bohne, L., Coquerelle, Th., Hagen, U.: Int. J. Radiat. Biol., in press (1970).

Boyce, R. P., Tepper, M.: Virology 34, 344 (1968).

Celano, A., Aurisicchio, S., Coppo, A., Domini, P., Graziosi, F.: Nuovo Cimento 18, 190 (1960).

Denhardt, D. T., Sinsheimer, R. L.: J. molec. Biol. 12, 663 (1965).

Dewey, D. L., Stein, G.: Nature 217, 351 (1968).

Freifelder, D.: Proc. nat. Acad. Sci. (Wash.) 54, 128 (1965).

— Virology 30, 328 (1966).

— Freifelder, D. R.: Mutation Res. 3, 177 (1966).

Ginoza, W.: Ann. Rev. Microbiol. 21, 325 (1967).
— Miller, R. C.: Proc. nat. Acad. Sci. (Wash.) 54, 551 (1965).
Harm, W.: In: Repair from genetic radiation damage. Ed.: F. H. Sobels. New York: Pergamon Press 1963a, p. 107.
— Z. Vererbungsl. 94, 67 (1963b).
— Mutation Res. 1, 344 (1964).
Haynes, R. H., Baker, R. M., Jones, G. E.: In: Energetics and mechanisms in radiation biology. Ed.: G. O. Phillips. London, New York: Academic Press 1968, p. 425.
Herak, J. N., Gordy, W.: Proc. nat. Acad. Sci. (Wash.) 56, 1354 (1966).
Hotz, G.: Biochem. biophys. Res. Commun. 11, 393 (1963).
— Z. Vererbungsl. 95, 211 (1964).
— Photochem. Photobiol., in press (1970).
— Reuschl, H.: Molec. Gen. Genetics 99, 5 (1967).
— Zimmer, K. G.: Int. J. Radiat. Biol. 7, 75 (1963).
Howard-Flanders, P.: Ann. Rev. Biochem. 37, 175 (1968).
— Boyce, R. P., Theriot, L.: Nature 195, 51 (1962).
Hüttermann, J., Müller, A.: Int. J. Radiat. Biol. 15, 297 (1969).
Ikenaga, M.: Radiat. Res. 34, 421 (1968).
Jung, H., Kürzinger, K.: Radiat. Res. 36, 369 (1968).
Kaplan, H. S.: In: Actions chimiques et biologiques des radiations, Tome 12. Ed.: M. Haissinsky. Paris: Masson et Cie 1968, p. 69.
Lytle, C. D., Ginoza, W.: Virology 38, 152 (1969a).
— — Int. J. Radiat. Biol. 14, 553 (1969b).
Müller, A., Köhnlein, W., Zimmer, K. G.: J. Mol. Biol. 7, 92 (1963).
Řeslová, S., Drobník, J.: Biochem. biophys. Res. Commun. 31, 119 (1968).
Rörsch, A., van de Camp, C., Adema, J.: Biochim. biophys. Acta (Amst.) 80, 246 (1964).
Rupert, C. S., Harm, W.: In: Advances in radiation biology, Vol. 2. Eds.: L. G. Augenstein, R. Mason, and M. Zelle. New York: Academic Press 1966, p. 1.
Sauerbier, W.: Virology 17, 164 (1962).
— Z. Vererbungsl. 95, 145 (1964a).
— Biochem. biophys. Res. Commun. 17, 46 (1964b).
Setlow, R., Boyce, R. P.: Biochim. biophys. Acta (Amst.) 68, 455 (1963).
Sinsheimer, R. L.: Progr. Nucleic Acid Res. 8, 115 (1968).
Stahl, F. W., Crasemann, J. M., Okun, L., Fox, E., Laird, C.: Virology 13, 98 (1961).
Stent, G. S.: Molecular biology of bacterial viruses. San Francisco: W. H. Freeman & Co 1963.
— Fuerst, C. R.: In: Advances in biological and medical physics, Vol. VII. Eds.: C. A. Tobias and J. H. Lawrence. New York: Academic Press 1960, p. 1.
Symonds, N.: J. molec. Biol. 4, 319 (1962).
Tanooka, H.: Radiat. Res. 21, 26 (1964).
Taylor, W. D., Ginoza, W.: Proc. nat. Acad. Sci. (Wash.) 58, 1753 (1967).
Tocchini-Valentini, G. P., Stodolsky, M., Aurisicchio, A., Sarnat, M., Graziosi, F., Weiss, S. B., Geiduschek, E. P.: Proc. nat. Acad. Sci. (Wash.) 50, 935 (1963).
Watson, J. D.: J. Bact. 60, 697 (1950).
— J. Bacteriol. 63, 473 (1952).
Weil, R.: Proc. nat. Acad. Sci. (Wash.) 49, 480 (1963).
Weiss, W., Richardson, C.: Proc. nat. Acad. Sci. (Wash.) 57, 1021 (1967).
Winkler, U.: Photochem. Photobiol. 3, 37 (1964a).
— Virology 24, 518 (1964b).
Yarus, M. J., Sinsheimer, R. L.: J. molec. Biol. 8, 614 (1964).

Chapter 13. The Action of Radiation on Bacteria

13.1. Some Basic Properties of Bacteria

Bacteria are the smallest, autonomous living systems with differentiated metabolic processes and able to reproduce independently. They also have a number of characteristics that make them a favoured radiation-biological cellular test system. They have a high metabolic activity and therefore reproduce rapidly, so that the effect of an irradiation can be observed after a relatively short period. For example, an E. coli bacterium divides about once in every 17 minutes under optimal growth conditions. Bacteria can be grown on well-defined media; if this is solid, *colonies* that can be counted with the naked eye are formed. The fact that the structure of the bacterial genome is very simple compared with that of higher cells is of particular interest for experiments in radiation biology and genetics. However, it is so simple that it might be considered to be a special case. In contrast to other cells, bacteria do not have a cell nucleus in the classical sense. Moreover, the division of the genetic material preceding normal cell division has few similarities with to the mechanism of chromosome duplication in higher cells; no spindle apparatus can be observed during this process. In spite of these phenomenological differences the terms nuclei and chromosomes are used, especially in the case of coli bacteria in which the genetic material gives the impression of being very compact. Bacteria contain varying numbers of nuclei, depending on their stage and conditions of growth.

Electron microscope investigations and autoradiographic studies have shown that the bacterial chromosomes consist of a double-stranded circular DNA molecule. In most cases, a "replicating arm" can be observed, which implies that DNA replication occurs during most of the cell cycle. E. coli DNA has the enormous molecular weight of about $3 \cdot 10^9$ Dalton.

The fact that bacteria not only divide by mitosis, but also show certain sexual characteristics, is of special importance in radiation biology. The phenomenon of *conjugation*, in which genetic material is transferred from a donor to an acceptor cell (F⁻) across a plasma bridge, makes it possible to carry out crossing experiments. The Hfr cells occupy a special position within the group of donor cells. They derive their name (Hfr = high frequency of recombination) from the fact that the progeny of the cross of Hfr to F⁻ have a high recombination frequency. Hfr cells of E. coli K12 can transfer their entire DNA linearly into the F⁻ cells within 89 minutes after scission of their genome ring, as long as the conjugation is not interrupted prior to completion (see Fig. 113). A bacterium becomes a donor through the action of what is known as *F factor*, or alternatively the *sex factor*, or the *F episome*. This is in the cytoplasm of F⁺ cells and replicates independently of the bacterial genome; in a manner of speaking, it is a "vagabond" gene. It can infect the F⁻ acceptor cell which does not contain the episome, causing it to become an F⁺ cell. If the F factor attaches itself to the genome, then the F⁺ cell becomes an Hfr cell. The F factor is not usually transferred in conjugation.

These characteristics of the F episome have some similarities to those of the *temperate phages*, which were encountered in Chapter 12.1. In fact, if the transformation considered in Chapter 11.3 is disregarded, then the transfer of genetic information between bacteria is in most cases based on an "episome-trick", which also occurs in the phenomenon of phage temperance. This is clearly shown by the example of the temperate phage λ, which transfers the galactose region ("gal") of the E. coli chromosome when it is induced after going through the prophage stage. This phenomenon is known as *transduction*. By analogy, the term *sexduction* refers to the transfer of a genetic marker by the F episome, after the prior passage through the Hfr stage. The F factor, after separating from the chromosome and channelling out of the cell, attaches itself in the newly-infected cells to the same site at which it was attached in the original cell, thus introducing a genetic marker from the previous Hfr "host".

These gene transfer processes, in particular Hfr crossing, can be used to form different types of mutants, especially those with differing radiation sensitivities. The very fact that there are such mutants is an indication of genetically-regulated repair processes, and therefore points directly to the DNA as the primary target for the action of radiation. The most common radiation-biological test on bacteria involves their inactivation, i. e. the loss of ability to form colonies. This test reflects their overall response to the action of radiation, as it develops from the primary lesion on the DNA to the subsequent failure of the cell. Within the framework of this chapter, conclusions about the primary processes in the DNA and the genetically-controlled repair of radiation lesions in DNA will be derived from inactivation tests. The most important details of those nucleic acid functions that are affected by the action of radiation have already been considered in Chapter 11. The absorption of radiation energy can also have a detrimental effect on bacterial *metabolism*. This aspect is, however, beyond the scope of this book.

13.2. Inactivation of Bacteria

Dose-response curves will be used as the starting point for the description and closer examination of inactivation. It cannot be concluded directly from these curves (as was possible with viruses) that DNA is the critical primary target, since too many modifying factors, such as growth conditions, pre- and post-irradiation treatments and particularly the capacity for repair of certain radiation lesions, have a pronounced influence on the shape of the dose-effect curves. It will therefore be necessary to demonstrate in some other way that the dose-response curves essentially reflect the sensitivity of DNA. If, however, it is initially assumed that DNA is the primary target, what conclusions can be drawn from the bacterial inactivation curves?

a) Ionizing Radiation

In a qualitative analysis of dose-response curves it is important to note that, in contrast to viruses, shoulder curves that approach exponential dependence asymptotically can be obtained with ionizing radiation. Fig. 100 shows, using the example of the E. coli mutant WP2 hcr⁺, that such shoulder curves are obtained with radiations of differing LET, although the shoulder becomes less pronounced at high LET. It is a significant feature of the curves in Fig. 100 that the asymptotic slope initially increases with LET, and then decreases again as the LET increases further. The considerations of Chapter 5.3 can, therefore, be applied to the asymptotic slopes of these curves at high doses, which are obviously a measure of radiation sensitivity. In a manner analogous to the procedure for exponential dose-effect curves, the quantity $1/D'_{37}$ is introduced for shoulder curves as a quantitative measure of radiation sensitivity. The D'_{37} is obtained by constructing a straight line through the origin of the coordinates, parallel to the asymptotic straight portion of the dose-response curve; the D'_{37} is then determined as in the case of single-hit curves. These factors have to be taken into account when bacterial inactivation curves are analyzed using hit and target theory. An obvious interpretation of shoulder curves is to consider them as single-hit multi-target curves. This type of curve has been discussed in detail in Chapter 2 (equation 2.6 and Fig. 5). Its characteristics are an asymptotic-exponential dependence and a finite extrapolation number equal to the number of targets (Fig. 5). Munson and Bridges (1966) chose this interpretation by identifying the extrapolation number with the number of chromatid strands per bacterium. This is not unreasonable since, depending on their stage of growth, bacteria may have several nuclei (chromosomes). In E. coli it is known that each chromosome is replicating during about 90% of the cell cycle. The average number of chromatid strands per chromosome is therefore about 1.44, and not 1. The extrapolation number is thus equal to the number of chromosomes per bacterium multiplied by 1.44. Munson and Bridges (1966) in experiments with certain E. coli B/r mutants found this to be approximately correct. The results are not, however, adequate to establish this concept, since such

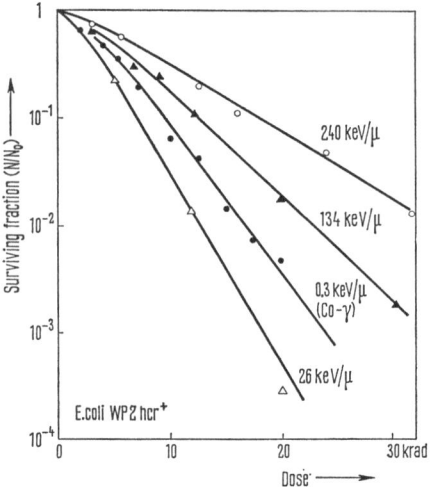

Fig. 100. Inactivation of E. coli WP2 hcr⁺ by ⁶⁰Co γ-radiation and α-particles with differing linear energy transfer (Bridges and Munson, 1968)

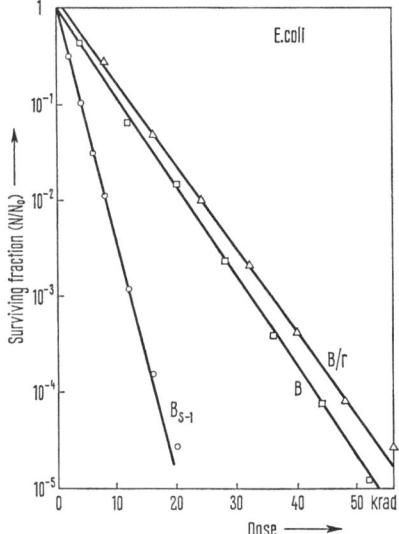

Fig. 101. Inactivation of various E. coli mutants by 150 kVp X-rays. (Haynes, 1964)

a simple interpretation seems to be unjustified in many cases, particularly as purely exponential inactivation curves are often obtained. Fig. 101 shows the exponential dose-response curves for the example of the wild type E. coli B and two mutants B/r and B_{s-1} having different radiation sensitivities. Despite the same DNA contents, curves with different slopes are obtained, particularly with B/r and B_{s-1}, due to their different repair capacities. There remains, therefore, the question whether the inactivation curves after irradiation could not be better described by a suitable stochastic expression, taking the mechanism of colony formation into account (see Chapter 3.5).

b) UV Light

A quite different situation seems to exist in the case of UV irradiation of bacteria, since different types of inactivation curves are obtained. Fig. 102 gives an example of this, showing the UV inactivation curve for the E. coli strains B, B/r, and B_{s-1}. As well as the pronounced shoulder curve of B/r, the high UV sensitivity of B_{s-1} is particularly noticeable; it decreases with increasing dose. The dose-response curve of the wild type E. coli B is also rather unusual. Despite the many differences between the curves, all three could have the same asymptotic slope. Since these three E. coli strains differ only in their capacity to repair radiation damage (in the wild type E. coli B, filamentous growth seems to play a part) an interpretation of UV inactivation could be considered in which the form of the curves is determined by repair capability at low and intermediate, but not at high doses. Haynes (1966) used this approach, describing the survival curve by the following general expression:

$$N/N_0 = e^{-[F(D)-R(D)]}, \qquad (13.1)$$

where $F(D)$ is the number of potentially-lethal lesions at a dose D, and $R(D)$ is the number of repaired lesions. The production of radiation lesions is proportional to dose

$$F(D) = k \cdot D. \qquad (13.2)$$

Haynes (1966) assumed the number of reparable lesions $R(D)$ to increase initially with dose, reaching a constant saturation at higher doses. The simplest mathematical form expressing this assumption is:

$$R(D) = \alpha(1 - e^{-\beta D}). \qquad (13.3)$$

Consequently, the inactivation curve (13.1) on a semi-logarithmic plot has an initial slope $k\text{-}\alpha\beta$, for $k > \alpha\beta$. The slope of this curve increases with dose and reaches a gradient k asymptotically. In this model the repair capability does not influence the final slope of this shoulder curve; instead the magnitude of the shoulder is a measure of the repaired UV lesions. The shoulder is characterized by the extrapolation number α, which represents the maximum number of repaired lesions. The extrapolation number for the resistant strain B/r obtained from Fig. 102 (Haynes, 1966) is $\alpha = 466$; i. e. in E. coli B/r, 466 UV-induced DNA lesions can be repaired, according to this model.

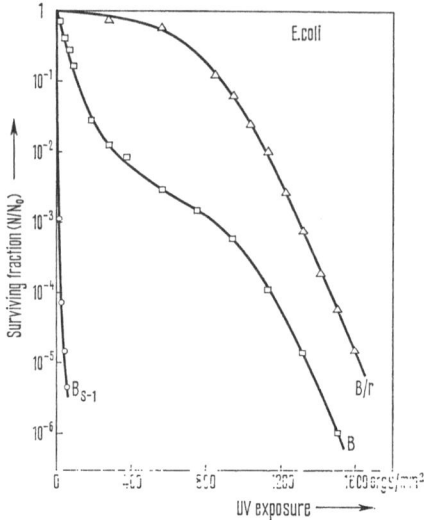

Fig. 102. Inactivation of various E. coli mutants by UV light of 254 nm wavelength. (Haynes, 1964)

A physico-chemical foundation for Haynes' repair hypothesis has been provided by the experiments of Hanawalt and Haynes (cited by Haynes *et al.*, 1968) which show that during the repair of UV-irradiated bacterial DNA, the incorporation of labelled thymine has a dose dependence which resembles the expression (13.3). This gives a considerable degree of support to the inactivation model described here, as will be shown in the following section in the discussion of the BU effect. However, these considerations cannot be applied in the case of ionizing radiation, since the capacity for repair then affects the gradient of the dose-response curve (Fig. 101). Before considering the repair of radiation lesions in detail, an attempt must be made to show that DNA is the critical target for inactivation, since the validity of the further arguments depends on this.

13.3. Bacterial DNA as the Critical Target

The emphasis will be concentrated on two types of investigation. Firstly, the influence of base composition on radiation sensitivity, and secondly the effect of incorporation of bromouracil into DNA.

a) The Influence of Base Composition

It has been shown that in many bacteria there is a correlation between the radiation sensitivity and the base composition of DNA expressed as the percentage adenine-thymine (A-T) content. Fig. 103 shows this correlation for several types of bacteria with UV as well as X-irradiation. The sensi-

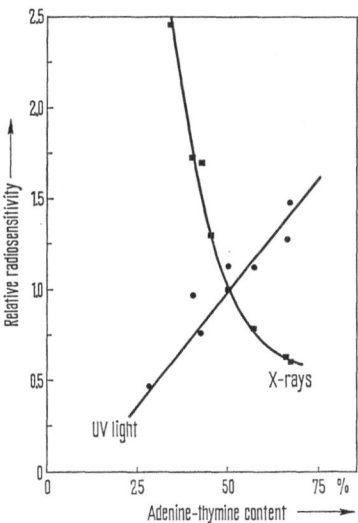

Fig. 103. Influence of the base composition of DNA on the sensitivity of bacteria to UV light and ionizing radiation. Adenine-thymine content of the bacteria common to both series: Ps. fluorescens 40%, S. marcescens 42%, E. coli 50%, B. subtilis 57%, M. pyogenes 66%, and B. cereus 67%. (Haynes, 1962; Kaplan and Zavarine, 1962)

tivity for 50% A-T (E. coli) was put equal to 1. An interesting observation is that the sensitivity to ionizing radiation decreases with increasing A-T content, while a significant increase in the sensitivity to UV light is observed. Care must be taken in the interpretation of these interesting observations, to which there are numerous exceptions. The fact should not be overlooked that radiation sensitivity certainly does not depend on base composition alone, but on numerous other parameters and on the repair capacity of the different strains. This is convincingly shown by the comparison of the radiation sensitivities of the different E. coli mutants (shown in Fig. 101 and 102), which with identical base compositions display large differences in their sensitivities. Strictly speaking, such an analysis of the radiation sensitivity of bacteria should begin with a division into groups which are similar with respect to the other parameters. This observation cannot, therefore, in itself be used as a conclusive argument for the assumption that DNA is the primary target for the action of radiation.

The increase in UV sensitivity with increasing A-T content can be attributed to the fact that the thymine dimer is the most important lethal lesion after UV irradiation, and that its rate of formation no doubt increases with increasing A-T content. It is not clear why the reverse dependence is observed with ionizing radiation. An indication of the possible reason for this correlation may lie in the nature of the radiation-induced base changes. Table 12 shows some identified radicals of DNA components, which were detected using electron spin resonance. It was found that in cytosine, in contrast to all the other bases, an attack occurs at the carbonyl group forming a COH radical (Dertinger, 1967), and probably ruptures the hydrogen bond on this site. The radical formation could result in local denaturation of the DNA which would explain the increase in the sensitivity to ionizing radiation that occurs with increasing cytosine-guanine content. It should be remembered, in this context, that with ionizing radiation the BU effect can also possibly be attributed to the formation of a COH radical (see Chapter 12.5), and that an increase in the radiation sensitivity with increasing bromouracil content of the DNA is also observed.

b) The BU Effect

A conclusive proof of the importance of DNA as the primary target in the action of radiation is provided by the incorporation of halogenated base analogues into bacterial DNA. Bromouracil (BU), which can be incorporated into DNA in place of thymine, is particularly important. Its sensitizing action on viral DNA has already been considered in Chapter 12.5; this chapter will therefore concentrate on the effects in bacterial DNA. An increase in sensitivity is observed in bacteria, as in viruses, when halogenated base analogues are incorporated into the DNA. The sensitivity increases with the degree of substitution, even when only one of the DNA strands contains the analogue (Kaplan *et al.*, 1962). Fig. 104 shows the influence of the maximal BU substitution on the *UV inactivation* curves of the thymine-deficient mutant E. coli BT⁻, into which BU can be relatively easily incorporated. It can be seen that although the incorporation of BU essentially removes the shoulder of the UV inactivation curve, the asym-

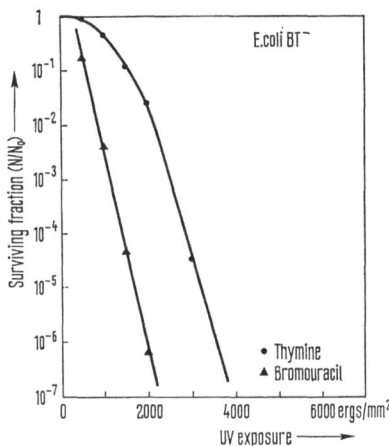

Fig. 104. UV inactivation of E. coli BT⁻ with normal and with bromouracil-substituted DNA. (Kaplan *et al.*, 1962)

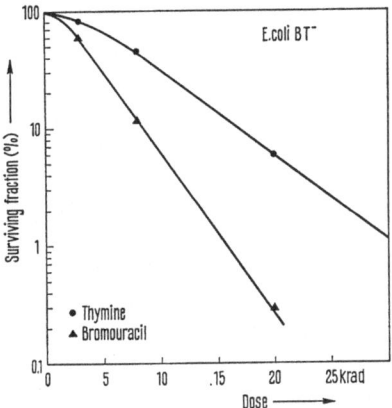

Fig. 105. Inactivation of E. coli BT⁻ with normal and bromouracil-substituted DNA by X-rays. (Kaplan *et al.*, 1962)

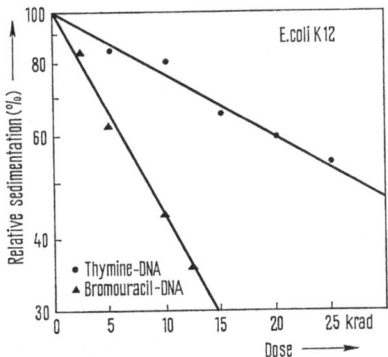

Fig. 106. Decrease in relative sedimentation of normal and bromouracil-substituted DNA from a E. coli K12 mutant in a neutral sucrose gradient after irradiation of the vegetative cells with ^{60}Co γ-rays. (Kaplan, 1966)

ptotic slope remains unchanged. The use of Haynes' inactivation hypothesis, which attributes the shoulder to the repair of UV damage, leads to the conclusion that the incorporation of BU in DNA prevents correct repair, as in the case of viruses (Chapter 12.5). This seems to be confirmed by the investigations of Aoki *et al.* (1966).

The situation is different with *ionizing radiation*, for the slope of the inactivation curve increases after incorporation of BU (Fig. 105). Kaplan (1966) showed that the increased sensitivity of bacteria containing BU-DNA can be correlated with the increase in the rate of DNA double strand breaks (Fig. 106). To show this, E. coli K 12 were lysed on a neutral sucrose gradient and the DNA sedimented. Under these conditions, double strand breaks can be detected. The relative sedimentation of DNA after BU

203

incorporation decreases with increasing radiation dose about 3 times as rapidly as that of normal DNA, as shown in Fig. 106. A similar sensitization effect was observed in an accompanying experiment when the colony forming ability was tested. In addition, Kaplan (1966) was able to show by sedimentation in alkaline gradients that repair of single strand breaks occurs in BU-DNA (an experiment of this kind is shown in Fig. 112).

Transforming DNA is sensitized to the same extent as colony formation by BU substitution (Szybalski and Opara-Kubinska, 1965). This fact, in conjunction with the other BU observations, shows that DNA is also the critical target in bacteria, and that radiation lesions in the DNA are responsible for the reproductive death of irradiated cells.

13.4. Repair of UV Damage

An important factor determining the radiation sensitivity of a cell is its ability to repair radiation damage in DNA. This repair capability has been examined in great detail in bacteria, so that today the mechanisms of some of these repair systems are on the whole understood. It is useful, as will become apparent in the following section, to consider the processes of repair of UV lesions before considering repair of those changes induced by ionizing radiation. The two most important mechanisms for the repair of UV lesions are photoreactivation and dark repair, both of which are enzymatically controlled, although they are basically different.

a) Photoreactivation

This is a highly efficient process, extending to many biological systems, as was already noted in early investigations of UV radiation effects. Indications of its existence can be traced back to 1904, but only after its rediscovery by Kelner in 1949 did this interesting process receive more attention. Photoreactivation, which also occurs in viruses (Chapter 12.4), is caused by an enzyme which in conjunction with light of wavelength between 300 and 400 nm eliminates lesions. In certain cases, photoreactivation processes may occur in the absence of an enzyme; these are known as *indirect photoreactivation* and are only mentioned in passing since the mechanism is not yet understood (Setlow, 1966). Enzymatic photoreactivation processes can take place in vitro, as is shown by Fig. 107 for the example of transforming DNA. When UV-irradiated DNA is incubated with a cell extract from E. coli B, the DNA is reactivated, which is reflected by an increase in the number of transformed cells with increasing incubation time. However, the transforming activity remains unaltered if the irradiated DNA is incubated with an extract from the mutant E. coli B phr⁻, which does not contain the photoreactivating enzymes. The following steps in the photoreactivation process have been demonstrated by such in vitro experiments:

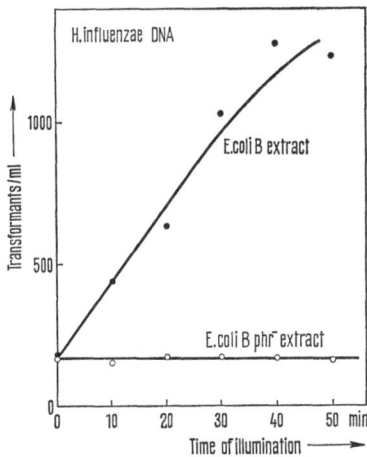

Fig. 107. In vitro photoreactivation of UV-irradiated transforming DNA of Haemophilus influenzae after incubation with extracts from E. coli B and E. coli B phr⁻ cells, respectively. (Rupert, 1965)

1. The enzyme forms a complex with UV-irradiated DNA (but not with unirradiated DNA) during incubation, even in the dark, as can be shown by centrifugation and column chromatography.

2. The main lesions that can be eliminated by exposure to light are the pyrimidine dimers, which are monomerized during the process of reactivation.

3. After exposure, the DNA-enzyme complex is split, and the repair process is complete.

The kinetics of intracellular complex formation can be followed by using single light flashes, about 1 millisecond in duration, and varying the time interval between UV irradiation and the flash reactivation (Harm et al., 1968). In stationary phase B_{s-1} cells irradiated with 4.8 ergs/mm² of 254 nm radiation (1 erg/mm² causes the formation of approximately 6.5 pyrimidine dimers per E. coli genome) the maximum number of complexes is formed within approximately 5 minutes at room temperature; 50⁰/₀ of the complexes are formed within the first 10 to 15 seconds. After greater UV doses, the maximum number of complexes reaches a constant limiting value of 20, indicating that this may be approximately the number of photoreactivating enzyme molecules in each cell. In some mutants of B_{s-1} this number may be as high as 100 molecules per cell (Harm, 1969). There are still some uncertainties about the details of the mechanisms by which light participates in this reactivation process, which can eliminate up to 90⁰/₀ of the UV lesions. For more general information the reader is referred to articles of Setlow (1966) and Rupert and Harm (1966). The site on the bacterial chromosome

205

of the regulating gene (phr) for the production of the photo-enzyme is shown in Fig. 113.

b) Dark Repair

This enzymatic process differs from photoreactivation in several aspects. First of all, repair may occur in the absence of light (as the name implies). The outstanding characteristic is, however, that not only are UV lesions (i. e. pyrimidine dimers) eliminated, but also, for example, cross-links formed by the action of alkylating agents such as nitrogen mustard, which links guanine bases on opposite strands of the DNA. The fact that such lesions, which obviously differ considerably from thymine dimers, can be repaired suggests that dark repair may be a universal correction process, which recognizes changes in the DNA configuration. In fact, there are a number of agents besides mustard gas the effects of which can be partially reversed by dark repair. Finally, dark repair differs from photoreactivation in that the cross-links and dimers are not merely split, but are excised by a complex process and replaced by newly-synthesized sections. This is the reason why dark repair is often also referred to as excision repair.

The fact that thymine dimers are excised from UV-irradiated DNA was discovered simultaneously by Setlow and Carrier (1964) and Boyce and Howard-Flanders (1964). These experiments, which cannot be discussed in detail, were carried out using UV-resistant and UV-sensitive mutants of E. coli, into the DNA of which ^{3}H-thymine had previously been incorporated. After UV irradiation, followed by incubation in the dark, the cells were lysed and the DNA precipitated with trichloroacetic acid. In the sensitive mutants, practically all of the radioactivity remained in the DNA, while in the resistant mutants, radioactive products derived from the excised DNA components were found in the supernant. Using paper chromatography, it was established that these components were thymine dimers, which occurred in the form of small oligonucleotides containing not more than three bases. These experiments demonstrated that the thymine dimer is not split in dark repair, but is excised as a whole.

In addition, the experiments of Pettijohn and Hanawalt (1964) showed that in the excision of a damaged site, resynthesis of the removed material occurs. The experiments were carried out using the deficient E. coli mutant TAU-bar which requires uracil, a series of amino acids and also thymine for growth. ^{3}H-thymine was incorporated into the bacterial DNA over a period of twelve generations. After UV irradiation, the cells were grown in a ^{14}C-BU containing medium. The BU was used in order to obtain a clear-cut sedimentation pattern of extracted DNA. The tritium activity, and therefore the quantity of thymine in DNA, decreases with increasing time of incubation in a BU-containing medium (as shown in Fig. 108) due to the excision of UV lesions, while the simultaneous increase in ^{14}C activity

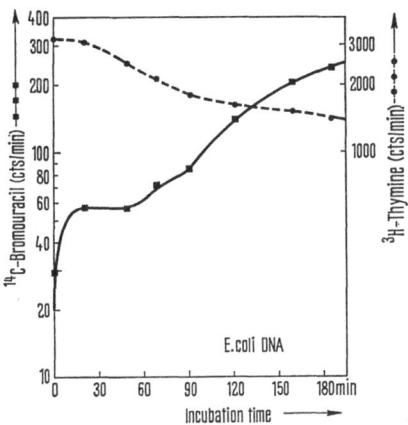

Fig. 108. Loss of ³H-thymine from and incorporation of ¹⁴C-bromouracil into the DNA of UV-irradiated E. coli cells as a function of the length of incubation of the cells after completion of irradiation. (Pettijohn and Hanawalt, 1964)

can be attributed to the incorporation of bromouracil during the reconstruction of the DNA. Pettijohn and Hanawalt concentrated on the proof of the fact that this BU is not incorporated during semi-conservative replication but is taken up during the repair of DNA. They extracted repaired DNA from the cells and centrifuged it in a CsCl gradient. It appeared in place of the "light" thymine-DNA on the sedimentation diagram, and could easily be identified by the radioactive BU label. This observation could only be interpreted in terms of excision of randomly distributed dimers from the DNA, and a subsequent replacement by BU; however, the total quantity of incorporated BU is so small that no measurable change in the sedimentation rate relative to "light" DNA is observed. In semi-conservatively replicated DNA, in contrast, the BU appears in the form of hybrid DNA (i. e. DNA which carries BU in a thymine-containing strand) as well as in the form of completely BU-substituted strands of DNA, and also in those DNA segments that were in replication. Degradation of DNA using ultrasonics or DNase gave further indications that bromouracil is incorporated into small sections of the DNA. These newly-synthesized sections contain at least 25, and possibly as many as several hundred nucleotides; i. e. not only thymine dimers are excised from the DNA strand, but also a number of nucleotides in the vicinity of such a lesion. It is interesting, although understandable in the light of the above discussion, that no BU is incorporated into DNA after a prior photo-reactivation, and that the UV-sensitive coli mutant B_{s-1} does not show this type of repair.

Two models of dark repair, based on the suggestions of Boyce and Howard-Flanders (1964) and Setlow and Carrier (1964), are in agreement

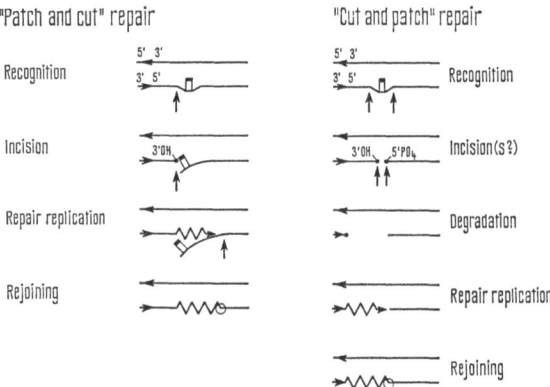

Fig. 109. Possible mechanisms for dark repair. (Haynes *et al.*, 1968)

with these results. They are known as the "patch and cut" and the "cut and patch" hypotheses (Hanawalt and Haynes, 1967). Both mechanisms, which are at present considered to be equally likely, are shown schematically in Fig. 109. In the cut and patch mechanism the following steps can be postulated:

1. Recognition of the lesion through a distortion of the DNA conformation.

2. Excision of the damage by a (double) endonuclease incision.

3. Widening of the gap by a 5', 3'-exonuclease, which is reflected in the observed degradation of DNA.

4. Repair synthesis with the help of a 3', 5'-polymerase, in which the intact complementary strand serves as a template.

5. Linking of the final phosphodiester bond by a polynucleotide-ligase.

The patch and cut model, shown on the left-hand side of Fig. 109, avoids the formation of a longer single-stranded section which is sensitive to shearing forces and therefore endangers the restitution of the DNA. After the endonuclease incision, repair synthesis occurs with the aid of a rather unusual polymerase which "peels" the defective strand until it is finally excised and degraded. A ligase forms the final diester bond. More information about the enzymatic processes involved in this repair is given in the review article by Howard-Flanders (1968).

The question still remains as to whether Haynes' inactivation hypothesis (Chapter 13.2) is in agreement with these models of dark repair. It is possible that the degradation process represents the limiting step that determines the maximum number of lesions repaired, since in the case of closely adjacent UV lesions (i. e. at higher doses) there is a danger that the DNA will be degraded to such an extent that reconstruction will no longer be possible. This assumption can be incorporated into a mathematical reformulation of the limited capacity for repair postulated by Haynes, and can also be used to calculate a dose-response curve under the condition that repair (and therefore survival) is only possible where the lesions are, on average, further apart than a certain critical distance. The adaptation of this formula to the survival of E. coli B/r (Fig. 102) gives a value of about 300 nucleotides for this critical distance. If this value is identified with the number of nucleotides excised by the action of exonuclease on a DNA segment, then it appears that in E. coli B/r the whole genetic marker containing the UV lesion is removed and replaced by newly-synthesized material during the repair process.

In this context it should be mentioned that not only UV-induced lethal lesions, but also lesions that lead to a mutation, can be eliminated by dark repair. Hill (1965) and Witkin (1966), for example, observed that certain mutations are induced at a higher rate by UV irradiation in repair-deficient coli strains, than in strains capable of carrying out repair.

13.5. Repair of Damage caused by Ionizing Radiation

The interesting question now arises of whether the mechanism of dark repair can be extended to damage induced by ionizing radiations, i. e. if it can repair single strand breaks in DNA. It can be shown by synergistic experiments using UV light and X-rays that the repair of X-ray induced damage may have steps in common with the repair of UV damage (Haynes, 1964). Fig. 110 shows that the slope of the X-ray inactivation curve for E. coli B/r increases after pre-exposure of the bacteria to various doses of UV light. The sensitivity of E. coli B/r is increased by this procedure, and finally reaches the approximate level of the sensitive E. coli B_{s-1}, which does not show this synergistic effect. On the other hand, X-ray pre-treatment only removes the shoulder on the inactivation curve of E. coli B/r, while the slope remains unaltered (Fig. 111). Thus, in both cases, pre-irradiation reduces the repair of the subsequent radiation-induced lesions, which implies that some of the reaction steps in the elimination of γ- and UV-induced radiation lesions are identical. This has already been suggested by Figs. 101 and 102, since E. coli B/r is more resistant than B_{s-1} to both UV and X-rays. However, this is not always the case. Bridges and Munson (1966) isolated a mutant E. coli WP2 hcr⁻ which is more sensitive to both UV and mustard gas than the hcr⁺ strain; however, both strains have the same sensitivities to the actions of X-rays and methyl-methane-sulphonate (MMS), an alkylating agent which causes DNA single breaks in vivo. Similar results have been obtained with B. subtilis by Searashi and Strauss (1965) and Reiter and Strauss (1965).

These observations show that the repair of single strand breaks can only be considered to be one part of the processes of dark repair, and that it may occur even when the complete dark repair of UV lesions is not possible; in other words, the UV repair appears to involve more "effort"

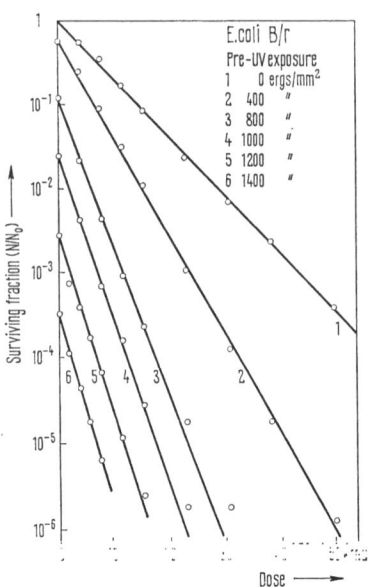

Fig. 110. Inactivation of E. coli B/r by X-rays after pre-irradiation with different UV doses. (Haynes, 1964)

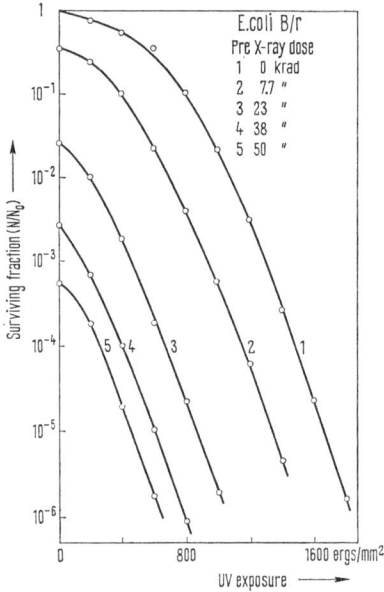

Fig. 111. UV inactivation of E. coli B/r after pre-irradiation with various doses of X-rays. (Haynes, 1964)

than repair of single breaks. Bacteria that are able to carry out a complete UV repair, can also repair single strand breaks; however, the converse is not necessarily true. As yet, no mutant that is resistant to UV but sensitive to X-rays has been found.

The occurrence of repair processes in bacteria irradiated with UV light or X-rays has different effects on the shape of the dose-response curves. While repair of UV lesions is reflected by a shoulder at small UV doses, repair of X-ray induced lesions influences the slope of the inactivation curves. Some steps of the reactions involved in the repair of X-ray lesions have been observed experimentally: for example, the degradation and subsequent resynthesis of DNA. McGrath et al. (1966) compared the coli mutants B/r and B_{s-1} with respect to the incorporation of ³H-thymidine into their DNA after irradiation, and to the release of radioactivity from the ³H-labelled DNA into the medium. Their results show, as expected, that B/r incorporated more ³H-thymidine than B_{s-1}. On the other hand, B_{s-1} releases more radioactivity into the medium than B/r. In B_{s-1}, DNA seems to be degraded rapidly upon X-irradiation, but with no reconstitution occurring. These findings may be taken as an indication that X-ray induced DNA lesions are exposed to different levels exonuclease activity in the various mutants having differing radiation sensitivities. Thus, the kinetics of repair in X-irradiated bacteria seem to reflect some kind of balance between degradation by exonuclease attack and repair synthesis (see also Chapter 13.6).

The repair of single strand breaks in the DNA of E. coli B/r has been followed directly by McGrath and Williams (1966). They used the sedimentation of DNA in an alkaline sucrose gradient, which allows the rate of single strand breaks to be determined. Their test system was the resistant strain E. coli B/r, the DNA of which contained ³H-thymidine. The sedimentation pattern obtained by lysis of the bacteria and centrifugation of the DNA is shown in Fig. 112. It is noticeable that a broad maximum occurs, even in the unirradiated control, since the E. coli DNA is already partially degraded by the preparative procedures (Fig. 112 a). Immediately after X-irradiation (dose: 20 krad) the maximum shifts to the right, and widens slightly (Fig. 112 b); i. e. the DNA single strands sediment more slowly, and in addition have a more heterogeneous weight distribution, which implies that fragments of varying lengths have been produced. After the incubation of B/r cells for 20 minutes at 37° C before the extraction of DNA, the maximum is marginally shifted towards the control value (Fig. 112 c). It regains its original position and shape only if the bacteria are incubated for 40 minutes before DNA extraction. Thus, most of the single strand breaks are repaired during incubation.

The hypothesis of the oxygen effect (Chapter 8.2) supplements many of the points considered in this section. Use has already been made of the

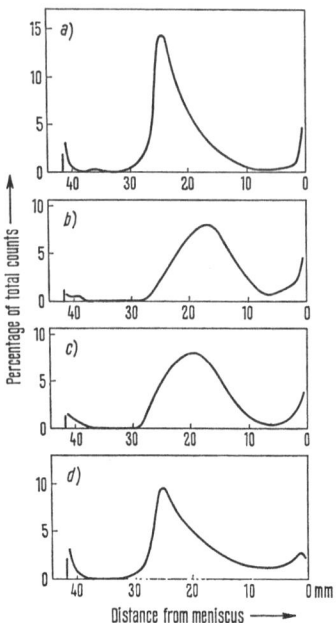

Fig. 112. Distribution of radioactivity in labelled DNA extracted from E. coli B/r after X-irradiation as a function of the distance from the meniscus after 90 minutes of centrifugation in an alkaline sucrose gradient. a. Unirradiated control. b. 20 krad, no incubation. c. 20 krad, followed by 20 minutes of incubation of the irradiated cells at 37° C. d. 20 krad, 40 minutes of incubation. (McGrath and Williams, 1966)

fact that the repair of radiation damage applies only to the Type 1 lesions, which refer essentially to single strand breaks and to base changes, i. e. lesions the production of which depends on the oxygen concentration. In contrast, the irreparable double strand breaks (Type 2) were assumed to be oxygen-independent, and this is confirmed by the results of Chapter 10.4.

13.6. Genetic Control of Repair in Bacterium E. coli

The repair of the radiation lesions has so far been considered as a specific process. Detailed investigations of the dark repair have, however, shown that repair of radiation lesions possibly represents a part of a general correction system which keeps the vital DNA functions unchanged, thereby ensuring the continuity of genetic information. In the following, various genetic factors influencing the radiation sensitivity and therefore the shape of inactivation curves are discussed. This study of different groups of E. coli mutants not only has the aim of constructing a gene map of the repair loci, but also has the intention of emphasizing the general aspects of repair and its relationship to other vital functions, e. g. to genetic recombination. More information relating to this section can be found in the articles of Howard-Flanders and Boyce (1966), Read (1968), Rörsch et al. (1967), and Taylor and Trotter (1967).

a) Mutants which form Filaments after UV or X-Irradiation

To this apparently unique group of mutants belong bacteria that show cell division after irradiation without, however, concomitant separation of the daughter cells. This leads to the formation of filamentous cells. The best known strain of this type is the wild type E. coli B; its higher radiation sensitivity relative to the B/r strain is attributed to its filament formation. It is therefore referred to as fil$^+$. On the other hand, irradiation of non-filament forming strains may lead to filament forming mutants, known as the dir$^-$ mutants ("division irradiation resistance"; Rörsch et al., 1967). The filament formation of dir$^-$ and fil$^+$ strains can be reduced by the addition of pantoyl-lactone. This substance probably splits the filament, thereby enabling the cells to form a colony. The dir$^-$ mutants correspond to the lon cells ("long form") of E. coli K12 strains (Howard-Flanders et al., 1964), which are also filamentous. The "lon" is probably a regulatory gene which influences the synthesis of the enzymes required for the formation of polysaccharides of the cell wall.

b) Mutants which cannot carry out Dark Repair of UV lesions, but are not Necessarily Sensitive to Ionizing Radiation

A particularly significant observation made by Howard-Flanders et al. (1966), was that dark repair is controlled by three regions of the bacterial genome, referred to as uvrA, B and C (there is possibly even a fourth gene). The uvr$^-$ mutants are very sensitive to UV radiation and the action of alkylating agents. However, uvr$^+$ and uvr$^-$ mutants are equally sensitive to those agents which produce single strand breaks, such as methyl-methane-sulphonate, and to ionizing radiation. Occasionally, uvr$^+$ mutants can repair a small fraction of X-ray damage, which may possibly consist of base changes. Multiple mutants, e. g. uvrA$^-$B$^-$, are only about 20% more UV sensitive than the single mutants, so that most probably the loci A, B and C do not merely regulate individual steps of dark repair. Most, but not all, of the uvr$^-$ mutants are unable to carry out host cell reactivation of bacteriophages. Rörsch and colleagues (1967) have described UV-sensitive coli mutants which are referred to as dar$^-$ ("dark repair"). Although most of these mutants belong to the groups A, B or C, they differ in their abilities to carry out host cell reactivation (Fig. 113), and also in other respects. The syn mutants ("synthesis of nucleic acids") which are part of the group uvrC (Rörsch et al., 1967) are also very UV sensitive. In syn$^-$ mutants, a considerable reduction in the synthesis of proteins and nucleic acids is observed after irradiation.

Thus, some of the mutants mentioned here can carry out host cell re-activation of phage DNA although their own DNA is not repaired. Al-though the significance of this phenomenon is not yet clear, it seems to support the supposition of Chapter 12.4, according to which the repair of phage DNA by the host cell is a simpler version of the dark repair of the cell's own DNA.

c) Mutants which are Sensitive to Both UV and X-Radiation

As already mentioned in Chapter 13.5, there are mutants that can repair neither UV nor X-ray damage. However, these must be double mutants since their sensitivity to X-rays, as has been shown, is basically independent of their repair capacity for UV damage. The corresponding loci responsible for X-ray sensitivity are referred to as exr ("X-ray resistant") or lex, and are mostly found in the uvrA region (Fig. 113). The best known double

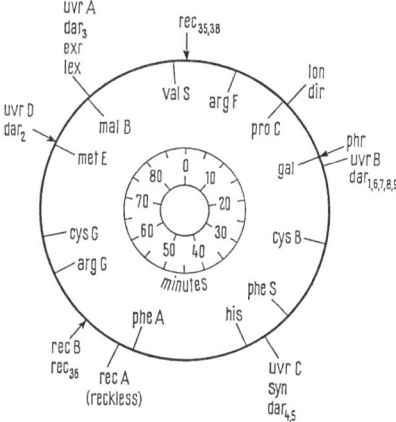

Fig. 113. E. coli chromosome with some genetic markers (on the inside) and the most important repair genes (on the outside), arranged as a function of the point of transfer into an acceptor cell after the start of conjugation in certain strains. Markers with positions not yet known accurately are indicated by arrows. (Rörsch et al., 1967; Taylor and Trotter, 1967; Read, 1968)

mutation of this kind exists between E. coli B_{s-1} and B/r, where B/r should be specified as uvr⁺ exr⁺ and B_{s-1} as uvr⁻ exr⁻. The exr mutation is in the uvrA group, and the uvr mutation in contrast is in the B region.

d) Mutants which cannot Undergo Genetic Recombination

The mutants, referred to as rec⁻, are very sensitive to both UV and X-rays. The defective recombination is reflected by the fact that no recombination occurs in the cross Hfr rec⁻×F⁻ rec⁻. However, the rec gene is dominant, i. e. recombination occurs if one of the partners has an intact rec gene. The fact that the rec⁻ mutants are very radiation sensitive should not be taken

214

as an indication that recombination is an important step in the repair process. It can represent, at the most, a partial overlap between the functions of repair and recombination, since the uvr-deficient mutation, in addition to the rec⁻ mutation, gives a further increase in the radiation sensitivity. Various observations with rec⁻ mutants lead to the conclusion that the rec gene (Fig. 113) is an exonuclease regulator, which allows the repair processes in rec⁻ mutants to get out of control to such an extent that the degradation of DNA becomes predominant. Consequently, various rec⁻ mutants show an uncontrolled, almost rampant, DNA degradation after irradiation, and are therefore referred to "reckless" mutants. They continuously lose nucleotides, even without being irradiated. There are also "cautious" rec⁻ mutants, which degrade DNA less avidly. This "regulator hypothesis" is supported by the observation that in rec⁻ mutants the activity levels of some enzymes acting on DNA do not differ from the levels in rec⁺ mutants (Clark et al., 1966; Gellert, 1967). It is interesting to note that the rec⁻ mutant normally has no influence on host cell reactivation, for example in T1-phages, which suggests that T1 possibly carries its own rec gene.

13.7. Micrococcus Radiodurans

All of the above considerations of radiation sensitivity and repair processes in bacteria seem to be contradicted by a phenomenon that is probably unique in the field of microorganisms; this is the radiation response of M. radiodurans, which owes its name to its almost legendary radiation resistance. Fig. 114 shows its X-ray inactivation curve under anaerobic conditions. It is characterized by an enormous shoulder, followed by an exponential fall-off. A D'_{37} of about 70 krad is derived from the slope of the exponential portion of the curve; this value is comparable with the D_{37} of 7 to 8 krad of the resistant E. coli B/r, since with due respect for the resistance of M. radiodurans, the fact should not be overlooked that the cell has only about 1/8th of the DNA content of a coli bacterium. The D'_{37} of M. radiodurans is, however, independent of oxygen pressure. This is rather unusual, and is contrary to observations with other bacteria where it is found that the oxygen enhancement increases with increasing radiation resistance (Chapter 8.3; Table 8). It must therefore be concluded that in the asymptotic portion of the curve in Fig. 114, M. radiodurans is inactivated only by double strand breaks, and that practically all other damage is repaired.

What, then, is the significance of the enormous shoulder observed both with X-rays (Fig. 114) and with UV radiation (Setlow, 1964)? Synergistic irradiation experiments, such as those already described for E. coli B/r (Fig. 110 and 111) do provide some information about this. However, the response of M. radiodurans to the combination of UV and X-rays is

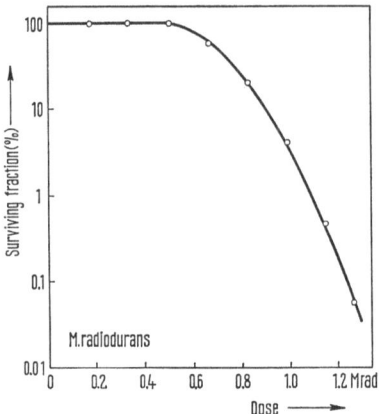

Fig. 114. Inactivation of Micrococcus radiodurans by X-rays. (Dean *et al.*, 1966)

fundamentally different from that of E. coli B/r. Moseley and Laser (1965) showed that there is no shoulder on the UV inactivation curve after pre-treatment with ionizing radiation, which is in agreement with the results obtained for E. coli B/r. Surprisingly, exactly the same effect is observed if the order of the irradiations is reversed. The final slope of the X-ray inactivation curve does not increase after prior UV irradiation, in contrast to E. coli B/r, but the shoulder does disappear. This implies that the kinetics of the repair of UV and X-ray damage are identical; in other words, the shoulder on the X-ray inactivation curve implies a "UV-like" repair of radiation damage. However, this formal analysis does not enable any conclusions to be made about the physico-chemical nature of the damage which is repaired in the region of the shoulder, but this is exactly the point where M. radiodurans is a genuine sensation: *it can repair double strand breaks in DNA.* This conclusion can be drawn from the form of the dose-response curve in Fig. 114. From the known G-values for the production of double strand breaks in DNA, it can be calculated that after a dose of 500 krad, an average of several double strand breaks must have been formed per DNA molecule, and yet a survival rate of practically 100% is registered. Repair of double strand breaks was first reported by Dean and colleagues (1966), who found numerous double strand breaks immediately after irradiation, which were recognizable by the reduction in viscosity of the DNA. This shows that the micrococcal DNA is no more resistant than other forms of double-stranded DNA. However, if the DNA was extracted after a two hour incubation, the viscosity was the same as that of control DNA, i. e. repair of double breaks must have occurred. It is, no doubt, aided by the fact that micrococcal DNA is surrounded by nucleoprotein, which prevents the broken DNA strands from separating. However, the mechanism of this repair of double strand breaks still remains a mystery. After UV or

216

X-irradiation, a release of DNA fragments is observed in M. radiodurans, which points to a mechanism similar to the excision repair observed in other resistant bacteria (Setlow, 1964; Lett et al., 1967). For example, the effectiveness with which the excision of thymine dimers occurs after UV irradiation led Jane K. Setlow (1964) to compare it with a "molecular striptease".

The chapter concludes with the discussion of this unusual system. The knowledge acquired here, and in the previous chapters (11 and 12) makes it possible to discuss, in a more general way, the connection between biological complexity and radiation sensitivity in the following final chapter. Those aspects which have crystallized as being essential will be used as guide lines, these being the critical target for radiation action, and the repair of radiation damage.

References

Aoki, S., Boyce, R. P., Howard-Flanders, P.: Nature 209, 686 (1966).
Boyce, R. P., Howard-Flanders, P.: Proc. nat. Acad. Sci. (Wash.) 51, 293 (1964).
Bridges, B. A., Munson, R. J.: Biochem. biophys. Res. Commun. 22, 268 (1966).
— — In: Current topics in radiation research, Vol. IV. Eds.: M. Ebert and A. Howard. Amsterdam: North-Holland Publ. Co. 1968, p. 95.
Clark, A. J., Chamberlin, M., Boyce, R. P., Howard-Flanders, P.: J. molec. Biol. 19, 442 (1966).
Dean, C. J., Feldschreiber, P., Lett, J. T.: Nature 209, 49 (1966).
Dertinger, H.: Z. Naturforsch. 22 b, 1266 (1967).
Gellert, M.: Proc. nat. Acad. Sci. (Wash.) 57, 148 (1967).
Hanawalt, P. C., Haynes, R. H.: Scient. American 216, 36 (1967).
Harm, W.: Mutation Res. 8, 411 (1969).
— Harm, H., Rupert, C. S.: Mutation Res. 6, 371 (1968).
Haynes, R. H.: Radiat. Res. 16, 562 (1962).
— Photochem. Photobiol. 3, 429 (1964).
— Radiat. Res. Suppl. 6, 1 (1966).
— Baker, R. M., Jones, G. E.: In: Energetics and mechanisms in radiation biology. Ed.: G. O. Phillips. London, New York: Academic Press 1968, p. 425.
Hill, R. F.: Photochem. Photobiol. 4, 563 (1965).
Howard-Flanders, P.: Ann. Rev. Biochem. 37, 175 (1968).
— Boyce, R. P.: Radiat. Res. Suppl. 6, 156 (1966).
— — Theriot, L.: Genetics 53, 1119 (1966).
— Simson, E., Theriot, L.: Genetics 49, 237 (1964).
Kaplan, H. S.: Proc. nat. Acad. Sci. (Wash.) 55, 1442 (1966).
— Smith, K. C., Tomlin, P. A.: Radiat. Res. 16, 98 (1962).
— Zavarine, R.: Biochem. biophys. Res. Commun. 8, 432 (1962).
Kelner, A.: Proc. nat. Acad. Sci. (Wash.) 35, 73 (1949).
Lett, J. T., Feldschreiber, P., Little, J. G., Steele, K., Dean, C. J.: Proc. roy. Soc. (Lond.) B 167, 184 (1967).
McGrath, R. A., Williams, R. W.: Nature 212, 534 (1966).
— — Swartzendruber, D. C.: Biophys. J. 6, 113 (1966).
Moseley, B. E. B., Laser, H.: Nature 206, 373 (1965).
Munson, R. J., Bridges, B. A.: Nature 210, 922 (1966).
Pettijohn, D., Hanawalt, P.: J. molec. Biol. 9, 395 (1964).

Read, J.: In: Actions chimiques et biologiques des radiations, Tome 12. Ed.: M. Haissinsky. Paris: Masson et Cie. 1968, p. 145.

Reiter, H., Strauss, B. S.: J. molec. Biol. 14, 179 (1965).

Rörsch, A., van de Putte, P., Mattern, I. E., Zwenk, H., van Sluis, C. A.: In: Radiation research. Ed.: G. Silini. Amsterdam: North-Holland Publ. Co. 1967, p. 771.

Rupert, C. S.: Photochem. Photobiol. 4, 271 (1965).

— Harm, W.: In: Advances in radiation biology, Vol. 2. Eds.: L. G. Augenstein, R. Mason, and M. Zelle. New York, London: Academic Press 1966, p. 1.

Searashi, T., Strauss, B. S.: Biochem. biophys. Res. Commun. 20, 680 (1965).

Setlow, J. K.: Photochem. Photobiol. 3, 405 (1964).

— Radiat. Res. Suppl. 6, 141 (1966).

Setlow, R. B., Carrier, W. L.: Proc. nat. Acad. Sci. (Wash.) 51, 226 (1964).

Szybalski, W., Opara-Kubinska, Z.: In: Cellular radiation biology. Baltimore: Williams and Wilkins 1965, p. 223.

Taylor, A. L., Trotter, C. D.: Bact. Rev. 31, 332 (1967).

Witkin, E. M.: Science 152, 1345 (1966).

Chapter 14. Radiation Sensitivity and Biological Complexity

With the investigation of the action of radiation on bacteria, the initially declared aims of this book have been achieved: to discuss the most important molecular mechanisms causing damage, and to describe their action on elementary biological systems. A certain basic idea has been repeatedly confirmed during the discussion of radiation sensitivity; this is the concept of the target theory, which basically states that the radiation sensitivity of a biological system increases with the size of its sensitive target. The target in enzymes was shown to be the whole molecule (see Fig. 28), and in viruses and bacteria the total DNA (see Chapters 12 and 13). However, when the examples of single and double strand viruses are considered, the relationship between radiation sensitivity $(1/D_{37})$ and the DNA molecular weight, requires the use of different proportionality constants, which are referred to as inactivation probabilities (killing-efficiency, MW_T/MW, Tables 15 and 16). The fact that different inactivation probabilities are obtained is due to the specific biological factors which decide whether or not a system can "survive" a specific lesion in the DNA, e. g. whether or not it can repair the lesion. It also, however, depends on the specific type of nucleic acid (single strand or double strand) and ultimately, if higher cells are included, on the specific arrangement of DNA in the chromosomes, which may even be present in multiples of the normal quantity (polyploidy).

14.1. Attempts at a Systematic Approach

The different radiation sensitivities of the various biological systems have led to numerous attempts to correlate radiation sensitivity with chemical and morphological characteristics, and thereby to clarify the nature of the inactivating events, as well as the type and size of the sensitive structure.

An approximate correlation with radiation sensitivity was noted in 1940 by Lea, and by Wollman and Lacassagne. They found that small viruses are more resistant to radiation than large ones. With better methods of preparation and detection which allowed exact measurements, Epstein found in 1953 that the radiation sensitivity of viruses correlates with their nucleic acid content rather than with their size; the relationship between DNA volume and radiation sensitivity was found to be approximately linear. A more recent presentation by Kaplan and Moses (1964) includes

219

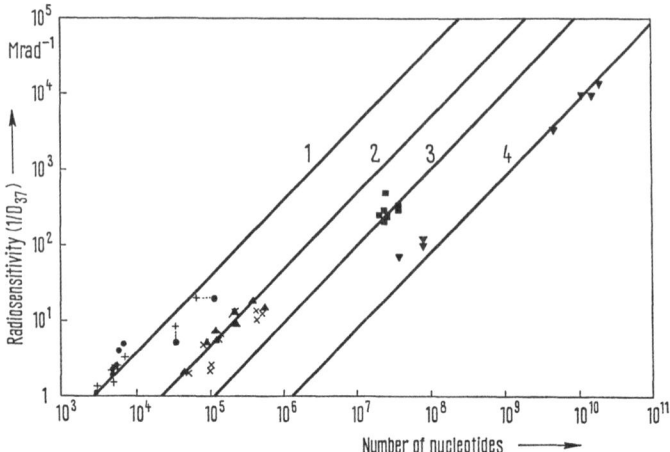

Fig. 115. Dependence of the radiation sensitivity of various biological objects on their nucleic acid content. Curve 1: Viruses with single-stranded DNA and RNA (+ values of Table 15). Curve 2: Viruses with double-stranded DNA (x values from Table 16). Curve 3: Haploid cells. Curve 4: Diploid cells. (Kaplan and Moses, 1964)

data not only on viruses, but also on bacteria, yeasts and cell cultures. Fig. 115 shows correlations between the radiation sensitivities ($1/D_{37}$) of different systems and their nucleic acid content (number of nucleotides). It was found that the experimental points could be described by four straight lines having a slope of 45°, where curve 1 represents viruses containing single-stranded RNA or DNA; curve 2, viruses containing double-stranded DNA; curve 3, haploid bacteria and yeast; and curve 4, diploid bacteria and yeasts, and avian and mammalian cells. This means that the experimental material can be divided into four classes which differ in their specific radiation sensitivities. Within these experimental classes, the increase in sensitivity is proportional to DNA content, which leads on a log-log plot to the observed straight lines of gradient 1 (45°). The inactivation probability in class 1 is approximately 1 (see Table 15), in class 2 approximately 0.1 (see Table 16), in class 3 approximately 0.02, and in class 4 of the order of magnitude of 0.002. Thus the radiation sensitivity at a constant nucleotide number decreases with increasing degree of biological organization in the systems analyzed by almost three orders of magnitude. While the reasons for the increased resistance in the transition from class 1 to class 2 have been extensively discussed in Chapter 12, it is not clear why there should be an increase in transition to class 3; since in addition to the fact that curve 3 (Fig. 115) is not defined unambiguously by the experimental points, it is plausible that all DNA lesions which are lethal for organisms in class 2 will also lead to inactivation of organisms in class 3. As it is known from Chapters 12 and 13 that repair of single strand breaks and base lesions in DNA occurs in bacteria and in double-stranded viruses, it is not quite clear why classes 2 and 3 do not coincide. Kaplan and Moses (1964) therefore investigated whether the slope of the four lines in Fig. 115 could be justified statistically. They showed that the experimental results are better described by four parallel lines with a gradient 0.809, than by lines with a gradient 1. The difference between the straight lines 2 and 3 are then no longer statistically significant, and the discussion of the reasons for the different inactivation probabilities is possibily rendered redundant. However, the assumption of a gradient of 0.809, which means that there is a power-dependence between radiation sensitivity and DNA content, has significant consequences which were discussed in detail by the authors.

In the transition from haploid to diploid cells, the radiation sensitivity decreases by a further factor of 10. This increase in resistance could be explained by improved repair processes in higher cells, especially since it is known that the DNA in mammalian cells is surrounded by protein, which may hold the broken DNA strands together until they are repaired. It is, however, just as possible that in the irradiation of haploid cells, recessive lethal mutations such as a point mutation or a small deletion may be responsible for a large proportion of the total inactivation observed, while in diploid organisms the inactivation is essentially due to dominant lethal mutations such as, for example, chromosome aberrations or breaks. In addition, it is known that not all of the DNA of such cells is necessarily required for divisions to occur, since occasionally part of a chromosome is lost during mitosis, but nevertheless a living progeny is formed ("redundancy of genetic information"). Finally, the fact that diploid cells contain two copies of each DNA molecule must be taken into consideration, as this may reduce their sensitivity even further. These arguments could explain why diploid cells are so resistant to ionizing radiation; however, the determination of the decisive factors will require a considerable extension of our knowledge of the mechanisms of DNA replication and cell division.

This scheme of Kaplan and Moses (1964) is not the only attempt that has been made to correlate radiation sensitivity with biological complexity in a multitude of different biological systems. The work of Terzi (1961, 1965) produced similar results, while Sparrow and colleagues (1967), by the inclusion of plant cells, arrived at eight different groups in which the radiation sensitivity is proportional to the volume of the interphase chromosomes. Of course, in summaries of this kind there is an inherent uncertainty which becomes particularly apparent when it is recalled that radiation sensitivity depends on the experimental conditions and can differ considerably for different mutants of the same system. The limitation of such summaries, and especially those covering a wide variety of different systems, must therefore be considered carefully; this is advisable in all target-theoretical considerations in any case, as has been repeatedly emphasized.

So far it has been assumed that *DNA is the critical target* in all cellular systems. This assumption is virtually confirmed by Fig. 115. It is also, however, possible to examine this important condition in higher cells. It was concluded in Chapter 11 that replication is the most sensitive function of DNA. This is reflected in the fact that radiation doses which almost completely inhibit cellular replication have practically no effect on non-dividing cells, e.g. nerve cells. The irradiation, with a sharply focussed proton beam, of mitotic cells of amphibian heart tissue cultures causes extensive chromosome aberrations when only a dozen protons have crossed the nucleus, while there is no reduction in the ability of the cells to divide after the passage of thousands of protons through the cytoplasm (Zirkle and Bloom, 1953). In this context, it is significant that the incorporation of BU into DNA increases the radiation sensitivity not only of viruses (see Chapter 12.5), and bacteria (see Chapter 13.3) but also of mammalian cells (Djordjevic and Szybalski, 1960). This evidence also supports the assumption of the dominant role of DNA as the critical target, the damage of which leads to reproductive death of the cell.

As well as DNA, the *cell membrane* could also be a structure that might be drastically changed by the occurrence of one or several energy absorption events, leading to the failure of the irradiated cell to reproduce. According to the hypothesis suggested by Bacq and Alexander (1961), enzymes are released after radiation-induced damage of membranes, and they then destroy important cellular structures by enzymatic attack. However, up until now, only degradation of DNA has been observed after irradiation, and this probably results mainly from the repair of radiation damage (see Chapter 13), while a significant degradation of other cell components cannot be found even after exposure to relatively high doses of radiation. Furthermore, it has been shown that the permeability of the cell membrane to a variety of substances only changes insignificantly, even after high radiation doses (see Bacq and Alexander, 1961). At this stage, there are therefore no experimental results that could be considered as proof that there are other targets in the cell that are of comparable significance to DNA.

14.2. What is Radiation Sensitivity?

To do full justice to the title of the final chapter of this book, a few remarks on the topic of radiation sensitivity are required. When referring to radiation sensitivity, the sensitivity of a specific molecular function is generally meant (such as the activity of an enzyme, or the DNA function of transformation), and not the radiation chemical sensitivity of the molecules. However, the aim of this book was to compare the radiation-chemical with the functional sensitivities. Although the results of these attempts can be seen in the relevant chapters, the résumé will be presented in a special form in order to give a better overall picture of the problem. In Fig. 116, the G-values for the destruction of a large number of molecules and macro-molecules, as well as the G-values for the inactivation of enzymes, viruses and microorganisms (the latter representing classes 1 to 4 of Fig. 115) are presented as a function of molecular weight (in the case of microorganisms, the molecular weight of their DNA). The experimental points in the region below 10^6 Dalton refer to water, simple organic compounds, dipeptides, and numerous enzymes. Most of the G-values of this group of compounds lie between 1 and 2, regardless of the molecular weight. Thus, the absorption of 50 to 100 eV of radiation energy in such a molecule always leads to its radiation chemical alteration, or inactivation. This value is in good agreement with the energy of 60 eV associated with a primary interaction (see Chapter 4.6); i. e. one primary ionization (ion cluster) results in the destruction of the molecule. In this region there is, therefore, *no true difference in resistance*, if the absorbed radiation energy *per molecule* is considered. This result is identical with the statement of target theory that the radiation sensitivities $(1/D_{37})$ of macromolecules are proportional to their molecular weights; since the dose is defined as absorbed energy *per unit mass* and not per molecule.

In the region of low molecular weights, the G-values shown in Fig. 116 increase by a factor of about 5 with decreasing molecular size. This increase in the number of altered molecules per 100 eV is not necessarily due to a higher sensitivity of smaller molecules; but rather, it reflects the fact that in the irradiation of small molecules, the ions and electrons belonging to the same ion cluster may affect different molecules, so that on average more than one molecule is changed per primary ionization (Hutchinson, 1966). With increasing molecular weight, it becomes increasingly probable that the ion pairs of a primary ionization will be localized within the same molecule. From this point of view, the G-values of 1 to 2 for macromolecules should not be interpreted as signifying that 50 to 100 eV are necessarily required for the inactivation of these molecules; this value represents the mean amount of energy deposited in a "not-too-small" volume by the interaction of ionizing radiation with matter.

There are also, however, examples of *true radiation resistance* on the molecular level. An example that should be noted is benzene, which due to its π-electron system is not only chemically stable, but also more resistant to ionizing radiation than non-aromatic compounds of the same size (Fig. 116). Apparently, conjugated ring systems are able to "redistribute" the absorbed radiation energy to several bonds so quickly that only rarely is a bond broken and the compound altered. This explanation is supported by the observation that an increase in the number of benzene rings per molecule increases the radiation resistance of a molecule (Fig. 116).

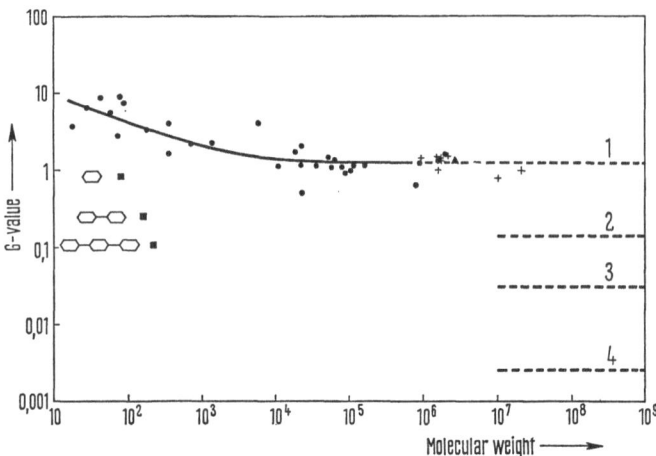

Fig. 116. G-values for the destruction or inactivation of molecules, macromolecules, and viruses as a function of their molecular weight or the molecular weight of their nucleic acids, respectively. ● Water, simple organic molecules, dipeptides, and enzymes. (Hutchinson, 1966). ▲ E. coli B ribosomes (Fig. 91). + Viruses with single-stranded nucleic acids (Table 15). ■ Benzene, diphenyl, and terphenyl (Burns and Jones, 1964). The broken lines 1, 2, 3, and 4 correspond to the straight lines in Fig. 115, where the formal G-values were calculated from equation (6.15) using the DNA molecular weights and the 37%-doses

The broken lines in Fig. 116 correspond to the straight lines drawn at 45° in Fig. 115. It is necessary, however, to be aware of the fact that a genuine G-value can be given only for systems whose genome consists of a single DNA molecule. In diploid cells, the "formal" G-value is only a measure of the inactivation probability; it should not be confused with the energy required for radiation chemical alteration of DNA. There are a variety of types of damage to the covalent structure of DNA, such as base changes, single strand breaks and double strand breaks (Chapter 10); if the G-values for these lesions are summed, using the results of Chapter 10, an average G-value lying between 1 and 2 is obtained for damage to DNA, which agrees well with the values for the other molecules shown in Fig. 116. The decrease in radiation sensitivity in the transition from class 1 to class 4 thus represents *no increase in the resistance of the DNA* in the systems concerned, but shows only that the functional sensitivity of the DNA decreases with increasing class number.

Thus the question which was the starting point for the interest in radiation biological investigation, i. e. why are cells inactivated by several hundred rads while enzymes, for example, can "tolerate" up to 100 Mrad, has been inverted. It is now necessary to explain why the organisms with increasing degrees of biological complexity are so resistant although they generally have a high DNA content. The reasons for this, in the cases of viruses and bacteria, have been discussed in detail in this book: the ability to repair radiation damage in the DNA was found to be one of the main reasons. At even higher degrees of organization, redundancy of genetic information and polyploidy may also play a part. Thus, the cellular effects of radiation increasingly show the character of a disturbance in cell function originating from the DNA, with the influence of metabolism playing an important part in the development of the disturbance. The description of *cellular radiation biology* therefore requires the metabolism of the cell to be taken into account to an increasing extent. This has not been necessary within the framework of *molecular radiation biology*, which concentrates on the basic physico-chemical mechanisms, and damage to DNA functions. The problems arising from cellular radiation biology extend, in turn, into the fields of *medical radiation biology* and *radiology*.

References

Bacq, Z. M., Alexander, P.: Fundamentals of radiobiology. Oxford: Pergamon Press 1961.
Burns, W. G., Jones, J. D.: Trans. Farad. Soc. 60, 2022 (1964).
Djordjevic, B., Szybalski, W.: J. exp. Med. 112, 509 (1960).
Epstein, H. T.: Nature 171, 394 (1953).
Hutchinson, F.: Cancer Res. 26, 2045 (1966).
Kaplan, H. S., Moses, L. E.: Science 145, 21 (1964).
Lea, D. E.: Nature 146, 137 (1940).
Sparrow, A. H., Underbrink, A. G., Sparrow, R. C.: Radiat. Res. 32, 915 (1967).
Terzi, M.: Nature 191, 461 (1961).
— J. theor. Biol. 8, 233 (1965).
Wollman, E., Lacassagne, A.: Ann. Inst. Pasteur 64, 5 (1940).
Zirkle, R. E., Bloom, W.: Science 117, 487 (1953).

Subject Index

226